Learn, Practice, Succeed

Eureka Math®
Grade 6
Module 1

Published by Great Minds®

Copyright © 2019 Great Minds®.

Printed in the U.S.A.

This book may be purchased from the publisher at eureka-math.org.

10   9   8   7   6   5   4   3   2

ISBN 978-1-64054-964-7

G6-M1-LPS-05.2019

Students, families, and educators:

Thank you for being part of the *Eureka Math*® community, where we celebrate the joy, wonder, and thrill of mathematics.

In *Eureka Math* classrooms, learning is activated through rich experiences and dialogue. That new knowledge is best retained when it is reinforced with intentional practice. The *Learn, Practice, Succeed* book puts in students' hands the problem sets and fluency exercises they need to express and consolidate their classroom learning and master grade-level mathematics. Once students learn and practice, they know they can succeed.

What is in the Learn, Practice, Succeed book?

Fluency Practice: Our printed fluency activities utilize the format we call a Sprint. Instead of rote recall, Sprints use patterns across a sequence of problems to engage students in reasoning and to reinforce number sense while building speed and accuracy. Sprints are inherently differentiated, with problems building from simple to complex. The tempo of the Sprint provides a low-stakes adrenaline boost that increases memory and automaticity.

Classwork: A carefully sequenced set of examples, exercises, and reflection questions support students' in-class experiences and dialogue. Having classwork preprinted makes efficient use of class time and provides a written record that students can refer to later.

Exit Tickets: Students show teachers what they know through their work on the daily Exit Ticket. This check for understanding provides teachers with valuable real-time evidence of the efficacy of that day's instruction, giving critical insight into where to focus next.

Homework Helpers and Problem Sets: The daily Problem Set gives students additional and varied practice and can be used as differentiated practice or homework. A set of worked examples, Homework Helpers, support students' work on the Problem Set by illustrating the modeling and reasoning the curriculum uses to build understanding of the concepts the lesson addresses.

Homework Helpers and Problem Sets from prior grades or modules can be leveraged to build foundational skills. When coupled with *Affirm*®, *Eureka Math*'s digital assessment system, these Problem Sets enable educators to give targeted practice and to assess student progress. Alignment with the mathematical models and language used across *Eureka Math* ensures that students notice the connections and relevance to their daily instruction, whether they are working on foundational skills or getting extra practice on the current topic.

Where can I learn more about Eureka Math *resources?*

The Great Minds® team is committed to supporting students, families, and educators with an evergrowing library of resources, available at eureka-math.org. The website also offers inspiring stories of success in the *Eureka Math* community. Share your insights and accomplishments with fellow users by becoming a *Eureka Math* Champion

Best wishes for a year filled with "aha" moments!

Jill Diniz

Jill Diniz
Chief Academic Officer, Mathematics
Great Minds

Contents

Module 1: Ratios and Unit Rates

Topic D: Percent

Example 1

The coed soccer team has four times as many boys on it as it has girls. We say the ratio of the number of boys to the number of girls on the team is $4 : 1$. We read this as *four to one*.

Suppose the ratio of the number of boys to the number of girls on the team is $3 : 2$.

Example 2: Class Ratios

Write the ratio of the number of boys to the number of girls in our class.

Write the ratio of the number of girls to the number of boys in our class.

Record a ratio for each of the examples the teacher provides.

1. _____   2. _____

3. _____   4. _____

5. _____   6. _____

Exercise 1

My own ratio compares _____ to

_____.

My ratio is _____.

Exercise 2

Using words, describe a ratio that represents each ratio below.

    a. 1 to 12 _____

               _____.

    b. $12 : 1$ _____

               _____.

    c. 2 to 5 _____

               _____.

d. 5 to 2 _____

_____.

e. 10 : 2 _____

_____.

f. 2 : 10 _____

_____.

> **Lesson Summary**
>
> A ratio is an ordered pair of numbers, which are not both zero.
>
> A ratio is denoted $A:B$ to indicate the order of the numbers—the number A is first, and the number B is second.
>
> The order of the numbers is important to the meaning of the ratio. Switching the numbers changes the relationship. The description of the ratio relationship tells us the correct order for the numbers in the ratio.

© 2019 Great Minds®. eureka-math.org

EUREKA MATH

Name _____   Date _____

1. Write a ratio for the following description: Kaleel made three times as many baskets as John during basketball practice.

2. Describe a situation that could be modeled with the ratio 4 : 1.

3. Write a ratio for the following description: For every 6 cups of flour in a bread recipe, there are 2 cups of milk.

1. At the local movie theatre, there are 115 boys, 92 girls, and 28 adults.

    a. Write the ratio of the number of boys to the number of girls.

    115 : 92

    b. Write the same ratio using another form ($A : B$ vs. A to B).

    115 to 92

    c. Write the ratio of the number of boys to the number of adults.

    115 : 28

    d. Write the same ratio using another form.

    115 to 28

> I know that I can represent a ratio using a colon or the word "to."

2. At a restaurant, 120 bottles of water are placed in ice at the buffet. At the end of the dinner rush, 36 bottles of water remained.

    a. What is the ratio of the number of bottles of water taken to the total number of water bottles?

    84 to 120, *or* 84 : 120

    b. What is the ratio of the number of water bottles remaining to the number of water bottles taken?

    36 to 84, or 36 : 84

> I need to subtract the number of water bottles remaining from the total number of water bottles to determine the number of water bottles taken.

3. Choose a situation that could be described by the following ratios, and write a sentence to describe the ratio in the context of the situation you chose.

    a. 1 to 3

    For every one meter, there are three feet.

    b. 7 to 30

    For every 7 days in a week, often there are 30 days in a month.

    c. 26 : 6

    For every 26 weeks, there are typically 6 months.

> I should choose situations that make sense with the numbers in the ratios. I know that for every one meter, there are three feet.

Lesson 1:   Ratios

1. At the sixth-grade school dance, there are 132 boys, 89 girls, and 14 adults.

    a. Write the ratio of the number of boys to the number of girls.

    b. Write the same ratio using another form ($A : B$ vs. A to B).

    c. Write the ratio of the number of boys to the number of adults.

    d. Write the same ratio using another form.

2. In the cafeteria, 100 milk cartons were put out for breakfast. At the end of breakfast, 27 remained.

    a. What is the ratio of the number of milk cartons taken to the total number of milk cartons?

    b. What is the ratio of the number of milk cartons remaining to the number of milk cartons taken?

3. Choose a situation that could be described by the following ratios, and write a sentence to describe the ratio in the context of the situation you chose.

    For example:

    $3 : 2$. When making pink paint, the art teacher uses the ratio $3 : 2$. For every 3 cups of white paint she uses in the mixture, she needs to use 2 cups of red paint.

    a. 1 to 2

    b. 29 to 30

    c. $52 : 12$

Exercise 1

Come up with two examples of ratio relationships that are interesting to you.

1.

2.

Exploratory Challenge

A T-shirt manufacturing company surveyed teenage girls on their favorite T-shirt color to guide the company's decisions about how many of each color T-shirt they should design and manufacture. The results of the survey are shown here.

Favorite T-Shirt Colors of Teenage Girls Surveyed

			X			
			X			
			X	X		
	X		X	X		X
	X		X	X	X	X
	X	X	X	X	X	X
X	X	X	X	X	X	X
Red	Blue	Green	White	Pink	Orange	Yellow

Exercises for Exploratory Challenge

1. Describe a ratio relationship, in the context of this survey, for which the ratio is $3:5$.

2. For each ratio relationship given, fill in the ratio it is describing.

Description of the Ratio Relationship (Underline or highlight the words or phrases that indicate the description is a ratio.)	Ratio
For every 7 white T-shirts they manufacture, they should manufacture 4 yellow T-shirts. The ratio of the number of white T-shirts to the number of yellow T-shirts should be ...	
For every 4 yellow T-shirts they manufacture, they should manufacture 7 white T-shirts. The ratio of the number of yellow T-shirts to the number of white T-shirts should be ...	
The ratio of the number of girls who liked a white T-shirt best to the number of girls who liked a colored T-shirt best was...	
For each red T-shirt they manufacture, they should manufacture 4 blue T-shirts. The ratio of the number of red T-shirts to the number of blue T-shirts should be ...	
They should purchase 4 bolts of yellow fabric for every 3 bolts of orange fabric. The ratio of the number of bolts of yellow fabric to the number of bolts of orange fabric should be ...	
The ratio of the number of girls who chose blue or green as their favorite to the number of girls who chose pink or red as their favorite was ...	
Three out of every 26 T-shirts they manufacture should be orange. The ratio of the number of orange T-shirts to the total number of T-shirts should be ...	

3. For each ratio given, fill in a description of the ratio relationship it could describe, using the context of the survey.

Description of the Ratio Relationship (Underline or highlight the words or phrases that indicate your example is a ratio.)	Ratio
	4 to 3
	3 : 4
	19 : 7
	7 to 26

Lesson 2:   Ratios

Lesson Summary

- Ratios can be written in two ways: A to B or $A : B$.

- We describe ratio relationships with words, such as *to, for each, for every*.

- The ratio $A : B$ is not the same as the ratio $B : A$ (unless A is equal to B).

Name _____   Date _____

Give two different ratios with a description of the ratio relationship using the following information:

There are 15 male teachers in the school. There are 35 female teachers in the school.

Examples

1. Using the design below, create 4 different ratios related to the image. Describe the ratio relationship, and write the ratio in the form $A : B$ or the form A to B.

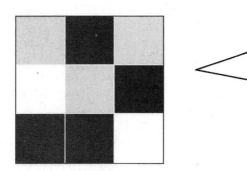

I see that there are 2 white tiles, 3 grey tiles, and 4 black tiles. I also see that there are 9 tiles altogether. I can use these quantities, the words "for each," "for every," or "to." I can also use a colon.

For every 9 tiles, there are 4 black tiles.

The ratio of the number of black tiles to the number of white tiles is 4 to 2.

The ratio of the number of grey tiles to the number of white tiles is 3 : 2.

There are 2 black tiles for each white tile.

Answers will vary.

2. Jaime wrote the ratio of the number of oranges to the number of pears as 2 : 3. Did Jaime write the correct ratio? Why or why not?

Jaime is incorrect. There are three oranges and two pears. The ratio of the number of oranges to the number of pears is 3 : 2.

I see that there are 3 oranges and 2 pears. I also know that the first value in the ratio relationship is the number of oranges, so that number is represented first in the ratio. The number of pears comes second in the relationship, so that number is represented second in the ratio.

1. Using the floor tiles design shown below, create 4 different ratios related to the image. Describe the ratio relationship, and write the ratio in the form $A : B$ or the form A to B.

2. Billy wanted to write a ratio of the number of apples to the number of peppers in his refrigerator. He wrote 1 : 3. Did Billy write the ratio correctly? Explain your answer.

Exercise 1

Write a one-sentence story problem about a ratio.

Write the ratio in two different forms.

Exercise 2

Shanni and Mel are using ribbon to decorate a project in their art class. The ratio of the length of Shanni's ribbon to the length of Mel's ribbon is $7 : 3$.

Draw a tape diagram to represent this ratio.

Exercise 3

Mason and Laney ran laps to train for the long-distance running team. The ratio of the number of laps Mason ran to the number of laps Laney ran was 2 to 3.

    a.   If Mason ran 4 miles, how far did Laney run? Draw a tape diagram to demonstrate how you found the answer.

    b.   If Laney ran 930 meters, how far did Mason run? Draw a tape diagram to determine how you found the answer.

    c.   What ratios can we say are equivalent to $2 : 3$?

Exercise 4

Josie took a long multiple-choice, end-of-year vocabulary test. The ratio of the number of problems Josie got incorrect to the number of problems she got correct is 2 : 9.

   a. If Josie missed 8 questions, how many did she get correct? Draw a tape diagram to demonstrate how you found the answer.

   b. If Josie missed 20 questions, how many did she get correct? Draw a tape diagram to demonstrate how you found the answer.

   c. What ratios can we say are equivalent to 2 : 9?

d. Come up with another possible ratio of the number Josie got incorrect to the number she got correct.

e. How did you find the numbers?

f. Describe how to create equivalent ratios.

> **Lesson Summary**
>
> Two ratios $A:B$ and $C:D$ are *equivalent ratios* if there is a nonzero number c such that $C = cA$ and $D = cB$. For example, two ratios are equivalent if they both have values that are equal.
>
> Ratios are equivalent if there is a nonzero number that can be multiplied by both quantities in one ratio to equal the corresponding quantities in the second ratio.

Name _____   Date _____

Pam and her brother both open savings accounts. Each begin with a balance of zero dollars. For every two dollars that Pam saves in her account, her brother saves five dollars in his account.

1. Determine a ratio to describe the money in Pam's account to the money in her brother's account.

2. If Pam has 40 dollars in her account, how much money does her brother have in his account? Use a tape diagram to support your answer.

3. Record the equivalent ratio.

4. Create another possible ratio that describes the relationship between the amount of money in Pam's account and the amount of money in her brother's account.

1. Write two ratios that are equivalent to $2 : 2$.

   $2 \times 2 = 4, 2 \times 2 = 4$; ***therefore, an equivalent ratio is*** $4 : 4$.

   $2 \times 3 = 6, 2 \times 3 = 6$; ***therefore, an equivalent ratio is*** $6 : 6$.

   Answers will vary.

   > The ratio is in the form $A : B$. I must multiply the A and B values by the same nonzero number to determine equivalent ratios.

2. Write two ratios that are equivalent to $5 : 13$.

   $5 \times 2 = 10, 13 \times 2 = 26$; ***therefore, an equivalent ratio is*** $10 : 26$.

   $5 \times 4 = 20, 13 \times 4 = 52$; ***therefore, an equivalent ratio is*** $20 : 52$.

3. The ratio of the length of the rectangle to the width of the rectangle is _____ to _____.

   > The length of this rectangle is 8 units, and the width is 5 units. Because the value for the length is listed first in the relationship, 8 is first in the ratio (or the A value). 5 is the B value.

   The ratio of the length of the rectangle to the width of the rectangle is $8 : 5$.

4. For a project in health class, Kaylee and Mike record the number of pints of water they drink each day. Kaylee drinks 3 pints of water each day, and Mike drinks 2 pints of water each day.

   a. Write a ratio of the number of pints of water Kaylee drinks to the number of pints of water Mike drinks each day.

   3 : 2

   b. Represent this scenario with tape diagrams.

   Number of pints of water Kaylee drinks

   Number of pints of water Mike drinks

   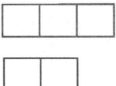

   c. If one pint of water is equivalent to 2 cups of water, how many cups of water did Kaylee and Mike each drink? How do you know?

   Kaylee drinks 6 cups of water because $3 \times 2 = 6$. Mike drinks 4 cups of water because $2 \times 2 = 4$. Since each pint represents 2 cups, I multiplied the number of pints of water Kaylee drinks by two and the number of pints of water Mike drinks by two. Also, since each unit represents two cups:

   Number of pints of water Kaylee drinks

   Number of pints of water Mike drinks

   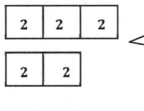

   Each unit in the tape diagrams represents 2 because there are two cups for every pint of water.

   d. Write a ratio of the number of cups of water Kaylee drinks to the number of cups of water Mike drinks.

   The ratio of the number of cups of water Kaylee drinks to the number of cups of water Mike drinks is 6 : 4.

   e. Are the two ratios you determined equivalent? Explain why or why not.

   3 : 2 and 6 : 4 are equivalent because they represent the same value. The diagrams never changed, only the value of each unit in the diagram.

Lesson 3:   Equivalent Ratios

EUREKA MATH

1. Write two ratios that are equivalent to 1 : 1.

2. Write two ratios that are equivalent to 3 : 11.

3.

   a. The ratio of the width of the rectangle to the height of the rectangle is _____ to _____.

   b. If each square in the grid has a side length of 8 mm, what is the width and height of the rectangle?

4. For a project in their health class, Jasmine and Brenda recorded the amount of milk they drank every day. Jasmine drank 2 pints of milk each day, and Brenda drank 3 pints of milk each day.

   a. Write a ratio of the number of pints of milk Jasmine drank to the number of pints of milk Brenda drank each day.

   b. Represent this scenario with tape diagrams.

   c. If one pint of milk is equivalent to 2 cups of milk, how many cups of milk did Jasmine and Brenda each drink? How do you know?

   d. Write a ratio of the number of cups of milk Jasmine drank to the number of cups of milk Brenda drank.

   e. Are the two ratios you determined equivalent? Explain why or why not.

Example 1

The morning announcements said that two out of every seven sixth-grade students in the school have an overdue library book. Jasmine said, "That would mean 24 of us have overdue books!" Grace argued, "No way. That is way too high." How can you determine who is right?

Exercise 1

Decide whether or not each of the following pairs of ratios is equivalent.

- If the ratios are not equivalent, find a ratio that is equivalent to the first ratio.
- If the ratios are equivalent, identify the nonzero number, c, that could be used to multiply each number of the first ratio by in order to get the numbers for the second ratio.

a. $6 : 11$ and $42 : 88$

_____ Yes, the value, c, is _____.

_____ No, an equivalent ratio would be _____.

b. $0 : 5$ and $0 : 20$

_____ Yes, the value, c, is _____.

_____ No, an equivalent ratio would be _____.

Exercise 2

In a bag of mixed walnuts and cashews, the ratio of the number of walnuts to the number of cashews is 5 : 6, Determine the number of walnuts that are in the bag if there are 54 cashews. Use a tape diagram to support your work. Justify your answer by showing that the new ratio you created of the number of walnuts to the number of cashews is equivalent to 5 : 6.

EUREKA
MATH

Lesson Summary

Recall the description:

Two ratios $A : B$ and $C : D$ are *equivalent ratios* if there is a positive number, c, such that $C = cA$ and $D = cB$. For example, two ratios are equivalent if they both have values that are equal.

Ratios are equivalent if there is a positive number that can be multiplied by both quantities in one ratio to equal the corresponding quantities in the second ratio.

This description can be used to determine whether two ratios are equivalent.

Name _____   Date _____

There are 35 boys in the sixth grade. The number of girls in the sixth grade is 42. Lonnie says that means the ratio of the number of boys in the sixth grade to the number of girls in the sixth grade is 5 : 7. Is Lonnie correct? Show why or why not.

Give two different ratios with a description of the ratio relationship using the following information:

There are 15 male teachers in the school. There are 35 female teachers in the school.

1. Use diagrams or the description of equivalent ratios to show that the ratios $4:5$, $8:10$, and $12:15$ are equivalent.

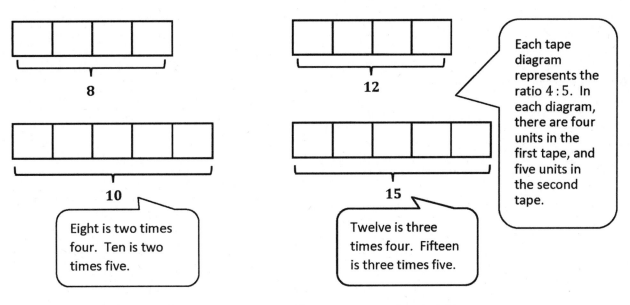

> Each tape diagram represents the ratio $4:5$. In each diagram, there are four units in the first tape, and five units in the second tape.

> Eight is two times four. Ten is two times five.

> Twelve is three times four. Fifteen is three times five.

The constant number, c, is 2.   *The constant number, c, is 3.*

2. The ratio of the amount of John's money to the amount of Rick's money is $5:13$. If John has \$25, how much money do Rick and John have together? Use diagrams to illustrate your answer.

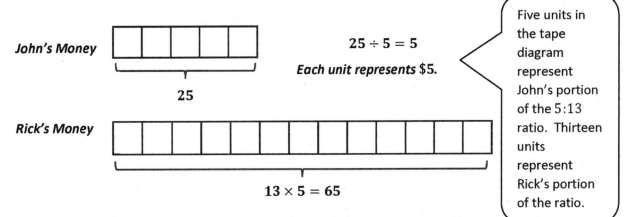

John's Money

$$25 \div 5 = 5$$

Each unit represents \$5.

25

> Five units in the tape diagram represent John's portion of the $5:13$ ratio. Thirteen units represent Rick's portion of the ratio.

Rick's Money

$$13 \times 5 = 65$$

5 units represents \$25. That means 1 unit represents \$5. Since all of the units are the same, 13 units represents \$65 because $13 \times 5 = 65$. *To determine how much money John and Rick have together, add the amounts.* $\$25 + \$65 = \$90$.

1. Use diagrams or the description of equivalent ratios to show that the ratios $2:3$, $4:6$, and $8:12$ are equivalent.

2. Prove that $3:8$ is equivalent to $12:32$.

    a. Use diagrams to support your answer.

    b. Use the description of equivalent ratios to support your answer.

3. The ratio of Isabella's money to Shane's money is $3:11$. If Isabella has \$33, how much money do Shane and Isabella have together? Use diagrams to illustrate your answer.

Example 1

A County Superintendent of Highways is interested in the numbers of different types of vehicles that regularly travel within his county. In the month of August, a total of 192 registrations were purchased for passenger cars and pickup trucks at the local Department of Motor Vehicles (DMV). The DMV reported that in the month of August, for every 5 passenger cars registered, there were 7 pickup trucks registered. How many of each type of vehicle were registered in the county in the month of August?

a.   Using the information in the problem, write four different ratios and describe the meaning of each.

b.   Make a tape diagram that represents the quantities in the part-to-part ratios that you wrote.

c.   How many equal-sized parts does the tape diagram consist of?

d.   What total quantity does the tape diagram represent?

e. What value does each individual part of the tape diagram represent?

f. How many of each type of vehicle were registered in August?

Example 2

The Superintendent of Highways is further interested in the numbers of commercial vehicles that frequently use the county's highways. He obtains information from the Department of Motor Vehicles for the month of September and finds that for every 14 non-commercial vehicles, there were 5 commercial vehicles. If there were 108 more noncommercial vehicles than commercial vehicles, how many of each type of vehicle frequently use the county's highways during the month of September?

Exercises

1. The ratio of the number of people who own a smartphone to the number of people who own a flip phone is $4:3$. If 500 more people own a smartphone than a flip phone, how many people own each type of phone?

2. Sammy and David were selling water bottles to raise money for new football uniforms. Sammy sold 5 water bottles for every 3 water bottles David sold. Together they sold 160 water bottles. How many did each boy sell?

3. Ms. Johnson and Ms. Siple were folding report cards to send home to parents. The ratio of the number of report cards Ms. Johnson folded to the number of report cards Ms. Siple folded is $2:3$. At the end of the day, Ms. Johnson and Ms. Siple folded a total of 300 report cards. How many did each person fold?

4. At a country concert, the ratio of the number of boys to the number of girls is $2:7$. If there are 250 more girls than boys, how many boys are at the concert?

Name _____   Date _____

When Carla looked out at the school parking lot, she noticed that for every 2 minivans, there were 5 other types of vehicles. If there are 161 vehicles in the parking lot, how many of them are not minivans?

1. The ratio of the number of females at a spring concert to the number of males is $7 : 3$. There are a total of 450 females and males at the concert. How many males are in attendance? How many females?

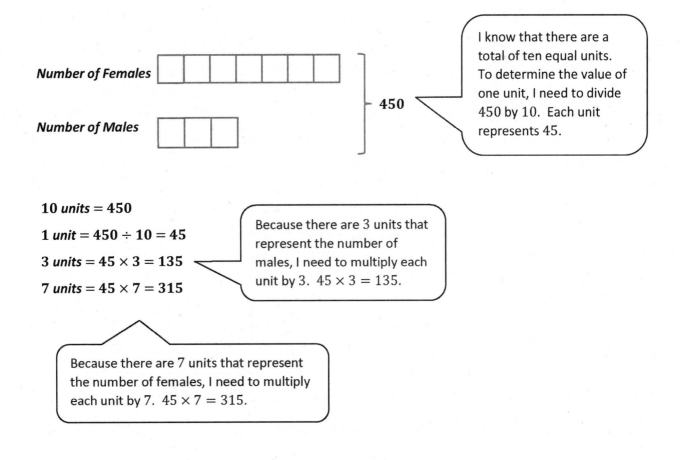

I know that there are a total of ten equal units. To determine the value of one unit, I need to divide 450 by 10. Each unit represents 45.

10 *units* $= 450$

1 *unit* $= 450 \div 10 = 45$

3 *units* $= 45 \times 3 = 135$

7 *units* $= 45 \times 7 = 315$

Because there are 3 units that represent the number of males, I need to multiply each unit by 3. $45 \times 3 = 135$.

Because there are 7 units that represent the number of females, I need to multiply each unit by 7. $45 \times 7 = 315$.

There are 135 *males and* 315 *females in attendance at the concert.*

2. The ratio of the number of adults to the number of students at a field trip has to be 3 : 8. During a current field trip, there are 190 more students on the trip than there are adults. How many students are attending the field trip? How many adults?

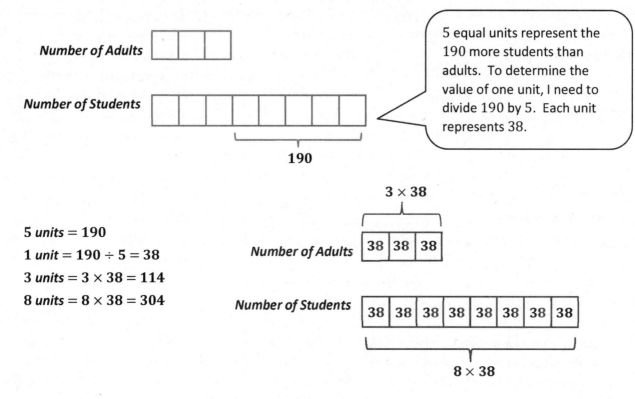

5 equal units represent the 190 more students than adults. To determine the value of one unit, I need to divide 190 by 5. Each unit represents 38.

5 *units* = 190
1 *unit* = 190 ÷ 5 = 38
3 *units* = 3 × 38 = 114
8 *units* = 8 × 38 = 304

There are 304 *students* and 114 *adults attending the field trip.*

Lesson 5:   Solving Problems by Finding Equivalent Ratios

1. Last summer, at *Camp Okey-Fun-Okey*, the ratio of the number of boy campers to the number of girl campers was 8 : 7. If there were a total of 195 campers, how many boy campers were there? How many girl campers?

2. The student-to-faculty ratio at a small college is 17 : 3. The total number of students and faculty is 740. How many faculty members are there at the college? How many students?

3. The Speedy Fast Ski Resort has started to keep track of the number of skiers and snowboarders who bought season passes. The ratio of the number of skiers who bought season passes to the number of snowboarders who bought season passes is 1 : 2. If 1,250 more snowboarders bought season passes than skiers, how many snowboarders and how many skiers bought season passes?

4. The ratio of the number of adults to the number of students at the prom has to be 1 : 10. Last year there were 477 more students than adults at the prom. If the school is expecting the same attendance this year, how many adults have to attend the prom?

Exercises

1. The Business Direct Hotel caters to people who travel for different types of business trips. On Saturday night there is not a lot of business travel, so the ratio of the number of occupied rooms to the number of unoccupied rooms is $2 : 5$. However, on Sunday night the ratio of the number of occupied rooms to the number of unoccupied rooms is $6 : 1$ due to the number of business people attending a large conference in the area. If the Business Direct Hotel has 432 occupied rooms on Sunday night, how many unoccupied rooms does it have on Saturday night?

2. Peter is trying to work out by completing sit-ups and push-ups in order to gain muscle mass. Originally, Peter was completing five sit-ups for every three push-ups, but then he injured his shoulder. After the injury, Peter completed the same number of repetitions as he did before his injury, but he completed seven sit-ups for every one push-up. During a training session after his injury, Peter completed eight push-ups. How many push-ups was Peter completing before his injury?

3. Tom and Rob are brothers who like to make bets about the outcomes of different contests between them. Before the last bet, the ratio of the amount of Tom's money to the amount of Rob's money was $4 : 7$. Rob lost the latest competition, and now the ratio of the amount of Tom's money to the amount of Rob's money is $8 : 3$. If Rob had \$280 before the last competition, how much does Rob have now that he lost the bet?

4. A sporting goods store ordered new bikes and scooters. For every 3 bikes ordered, 4 scooters were ordered. However, bikes were way more popular than scooters, so the store changed its next order. The new ratio of the number of bikes ordered to the number of scooters ordered was $5 : 2$. If the same amount of sporting equipment was ordered in both orders and 64 scooters were ordered originally, how many bikes were ordered as part of the new order?

5. At the beginning of Grade 6, the ratio of the number of advanced math students to the number of regular math students was $3 : 8$. However, after taking placement tests, students were moved around changing the ratio of the number of advanced math students to the number of regular math students to $4 : 7$. How many students started in regular math and advanced math if there were 92 students in advanced math after the placement tests?

6. During first semester, the ratio of the number of students in art class to the number of students in gym class was 2 : 7. However, the art classes were really small, and the gym classes were large, so the principal changed students' classes for second semester. In second semester, the ratio of the number of students in art class to the number of students in gym class was 5 : 4. If 75 students were in art class second semester, how many were in art class and gym class first semester?

7. Jeanette wants to save money, but she has not been good at it in the past. The ratio of the amount of money in Jeanette's savings account to the amount of money in her checking account was 1 : 6. Because Jeanette is trying to get better at saving money, she moves some money out of her checking account and into her savings account. Now, the ratio of the amount of money in her savings account to the amount of money in her checking account is 4 : 3. If Jeanette had $936 in her checking account before moving money, how much money does Jeanette have in each account after moving money?

> **Lesson Summary**
>
> When solving problems in which a ratio between two quantities changes, it is helpful to draw a *before* tape diagram and an *after* tape diagram.

Name _____   Date _____

Students surveyed boys and girls separately to determine which sport was enjoyed the most. After completing the boy survey, it was determined that for every 3 boys who enjoyed soccer, 5 boys enjoyed basketball. The girl survey had a ratio of the number of girls who enjoyed soccer to the number of girls who enjoyed basketball of 7 : 1. If the same number of boys and girls were surveyed, and 90 boys enjoy soccer, how many girls enjoy each sport?

Solving Ratio Problems

At the beginning of Grade 6, the ratio of the number of students who chose art as their favorite subject to the number of students who chose science as their favorite subject was $4:9$. However, with the addition of an exciting new art program, some students changed their mind, and after voting again, the ratio of the number of students who chose art as their favorite subject to the number of students who chose science as their favorite subject changed to $6:7$. After voting again, there were 84 students who chose art as their favorite subject. How many fewer students chose science as their favorite subject after the addition of the new art program than before the addition of the new art program? Explain.

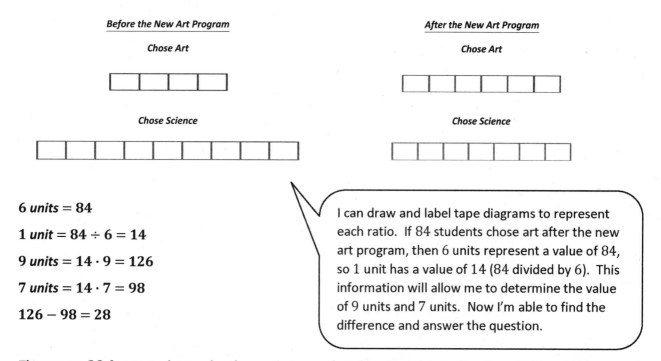

6 units = 84

1 unit = 84 ÷ 6 = 14

9 units = 14 · 9 = 126

7 units = 14 · 7 = 98

126 − 98 = 28

I can draw and label tape diagrams to represent each ratio. If 84 students chose art after the new art program, then 6 units represent a value of 84, so 1 unit has a value of 14 (84 divided by 6). This information will allow me to determine the value of 9 units and 7 units. Now I'm able to find the difference and answer the question.

There were 28 fewer students who chose science as their favorite subject after the addition of the new art program than the number of students who chose science as their favorite subject before the addition of the new art program. 126 students chose science before, and 98 students chose science after the new art program was added.

1. Shelley compared the number of oak trees to the number of maple trees as part of a study about hardwood trees in a woodlot. She counted 9 maple trees to every 5 oak trees. Later in the year there was a bug problem and many trees died. New trees were planted to make sure there was the same number of trees as before the bug problem. The new ratio of the number of maple trees to the number of oak trees is $3 : 11$. After planting new trees, there were 132 oak trees. How many more maple trees were in the woodlot before the bug problem than after the bug problem? Explain.

2. The school band is comprised of middle school students and high school students, but it always has the same maximum capacity. Last year the ratio of the number of middle school students to the number of high school students was $1 : 8$. However, this year the ratio of the number of middle school students to the number of high school students changed to $2 : 7$. If there are 18 middle school students in the band this year, how many fewer high school students are in the band this year compared to last year? Explain.

Example 1

Which of the following correctly models that the number of red gumballs is $\frac{5}{3}$ the number of white gumballs?

a. Red □□□
    White □□□□□

b. Red □□□□□
    White □□

c. Red □□□
    White □□

d. Red □□□□□
    White □□□□□□□□

Example 2

The duration of two films are modeled below.

Film A □□□□□

Film B □□□□□□□

a. The ratio of the length of Film A to the length of Film B is _____ : _____.

b. The length of Film A is $\dfrac{\square}{\square}$ of the length of Film B.

c. The length of Film B is $\dfrac{\square}{\square}$ of the length of Film A.

Exercise 1

Sammy and Kaden went fishing using live shrimp as bait. Sammy brought 8 more shrimp than Kaden brought. When they combined their shrimp they had 32 shrimp altogether.

    a.   How many shrimp did each boy bring?

    b.   What is the ratio of the number of shrimp Sammy brought to the number of shrimp Kaden brought?

    c.   Express the number of shrimp Sammy brought as a fraction of the number of shrimp Kaden brought.

    d.   What is the ratio of the number of shrimp Sammy brought to the total number of shrimp?

    e.   What fraction of the total shrimp did Sammy bring?

     Lesson 7:   Associated Ratios and the Value of a Ratio

Exercise 2

A food company that produces peanut butter decides to try out a new version of its peanut butter that is extra crunchy, using twice the number of peanut chunks as normal. The company hosts a sampling of its new product at grocery stores and finds that 5 out of every 9 customers prefer the new extra crunchy version.

    a.   Let's make a list of ratios that might be relevant for this situation.

        i.   The ratio of number preferring new extra crunchy to total number surveyed is _____.

        ii.   The ratio of number preferring regular crunchy to the total number surveyed is _____.

        iii.   The ratio of number preferring regular crunchy to number preferring new extra crunchy is _____.

        iv.   The ratio of number preferring new extra crunchy to number preferring regular crunchy is _____.

    b.   Let's use the value of each ratio to make multiplicative comparisons for each of the ratios we described here.

        i.   The number preferring new extra crunchy is _____ of the total number surveyed.

        ii.   The number preferring regular crunchy is _____ of the total number surveyed.

        iii. The number preferring regular crunchy is _____ of those preferring new extra crunchy.

        iv.   The number preferring new extra crunchy is _____ of those preferring regular crunchy.

    c.   If the company is planning to produce 90,000 containers of crunchy peanut butter, how many of these containers should be the new extra crunchy variety, and how many of these containers should be the regular crunchy peanut butter? What would be helpful in solving this problem? Does one of our comparison statements above help us?

Try these next scenarios:

d. If the company decides to produce 2,000 containers of regular crunchy peanut butter, how many containers of new extra crunchy peanut butter would it produce?

e. If the company decides to produce 10,000 containers of new extra crunchy peanut butter, how many containers of regular crunchy peanut butter would it produce?

f. If the company decides to only produce 3,000 containers of new extra crunchy peanut butter, how many containers of regular crunchy peanut butter would it produce?

Lesson Summary

For a ratio $A : B$, we are often interested in the associated ratio $B : A$. Further, if A and B can both be measured in the same unit, we are often interested in the associated ratios $A : (A + B)$ and $B : (A + B)$.

For example, if Tom caught 3 fish and Kyle caught 5 fish, we can say:

The ratio of the number of fish Tom caught to the number of fish Kyle caught is $3 : 5$.

The ratio of the number of fish Kyle caught to the number of fish Tom caught is $5 : 3$.

The ratio of the number of fish Tom caught to the total number of fish the two boys caught is $3 : 8$.

The ratio of the number of fish Kyle caught to the total number of fish the two boys caught is $5 : 8$.

For the ratio $A : B$, where $B \neq 0$, the value of the ratio is the quotient $\frac{A}{B}$.

For example: For the ratio $6 : 8$, the value of the ratio is $\frac{6}{8}$ or $\frac{3}{4}$.

Name _____ Date _____

Alyssa's extended family is staying at the lake house this weekend for a family reunion. She is in charge of making homemade pancakes for the entire group. The pancake mix requires 2 cups of flour for every 10 pancakes.

1. Write a ratio to show the relationship between the number of cups of flour and the number of pancakes made.

2. Determine the value of the ratio.

3. Use the value of the ratio to fill in the following two multiplicative comparison statements.

    a. The number of pancakes made is _____ times the amount of cups of flour needed.

    b. The amount of cups of flour needed is _____ of the number of pancakes made.

4. If Alyssa has to make 70 pancakes, how many cups of flour will she have to use?

1. Amy is making cheese omelets for her family for breakfast to surprise them. For every 2 eggs, she needs $\frac{1}{2}$ cup of cheddar cheese. To have enough eggs for all the omelets she is making, she calculated she would need 16 eggs. If there are 5.5 cups of cheddar cheese in the fridge, does Amy have enough cheese to make the omelets? Why or why not?

The ratio of the number eggs to the number of cups of cheese is $2:\frac{1}{2}$.

The value of the ratio is 4.

> I need to determine the value of the ratio in order to find the amount of cheese that is needed. I can do this by dividing 2 by $\frac{1}{2}$. The number of cups of cheese needed is $\frac{1}{4}$ the number of eggs. I can also say the number of eggs is 4 times the number of cups of cheese.

$$2:\frac{1}{2} \qquad\qquad 16:4$$

2 is four times as much as $\frac{1}{2}$.

16 is four times as much as 4.

Amy needs 4 cups of cheddar cheese to make the omelets with 16 eggs. She will have enough cheese because she needs 4 cups and has 5.5 cups.

2. Samantha is a part of the Drama Team at school and needs pink paint for a prop they're creating for the upcoming school play. Unfortunately, the 6 gallons of pink paint she bought is too dark. After researching how to lighten the paint to make the color she needs, she found out that she can mix $\frac{1}{3}$ of a gallon of white paint with 2 gallons of the pink paint she bought. How many gallons of white paint will Samantha have to buy to lighten the 6 gallons of pink paint?

The ratio of the number of gallons of white paint to the number of gallons of pink paint is $\frac{1}{3}:2$.

The value of the ratio is $\frac{1}{6}$.

> I need to determine the value of the ratio by dividing $\frac{1}{3}$ by 2. The number of gallons of white paint needed is $\frac{1}{6}$ of the number of gallons of pink paint. I can also say the number of gallons of pink paint is 6 times the number of gallons of white paint.

$\frac{1}{3}$ is $\frac{1}{6}$ of 2.

1 is $\frac{1}{6}$ of 6.

Samantha would need 1 gallon of white paint to make the shade of pink she desires.

1. Maritza is baking cookies to bring to school and share with her friends on her birthday. The recipe requires 3 eggs for every 2 cups of sugar. To have enough cookies for all of her friends, Maritza determined she would need 12 eggs. If her mom bought 6 cups of sugar, does Maritza have enough sugar to make the cookies? Why or why not?

2. Hamza bought 8 gallons of brown paint to paint his kitchen and dining room. Unfortunately, when Hamza started painting, he thought the paint was too dark for his house, so he wanted to make it lighter. The store manager would not let Hamza return the paint but did inform him that if he used $\frac{1}{4}$ of a gallon of white paint mixed with 2 gallons of brown paint, he would get the shade of brown he desired. If Hamza decided to take this approach, how many gallons of white paint would Hamza have to buy to lighten the 8 gallons of brown paint?

Exercise 1

Circle any equivalent ratios from the list below.

      Ratio: $1:2$

      Ratio: $5:10$

      Ratio: $6:16$

      Ratio: $12:32$

Find the value of the following ratios, leaving your answer as a fraction, but rewrite the fraction using the largest possible unit.

      Ratio: $1:2$ Value of the Ratio:

      Ratio: $5:10$ Value of the Ratio:

      Ratio: $6:16$ Value of the Ratio:

      Ratio: $12:32$ Value of the Ratio:

What do you notice about the value of the equivalent ratios?

Exercise 2

Here is a theorem: If $A:B$ with $B \neq 0$ and $C:D$ with $D \neq 0$ are equivalent, then they have the same value: $\dfrac{A}{B} = \dfrac{C}{D}$.

This is essentially stating that if two ratios are equivalent, then their values are the same (when they have values).

Can you provide any counterexamples to the theorem above?

Exercise 3

Taivon is training for a duathlon, which is a race that consists of running and cycling. The cycling leg is longer than the running leg of the race, so while Taivon trains, he rides his bike more than he runs. During training, Taivon runs 4 miles for every 14 miles he rides his bike.

    a.   Identify the ratio associated with this problem and find its value.

Use the value of each ratio to solve the following.

    b.   When Taivon completed all of his training for the duathlon, the ratio of total number of miles he ran to total number of miles he cycled was 80 : 280. Is this consistent with Taivon's training schedule? Explain why or why not.

    c.   In one training session, Taivon ran 4 miles and cycled 7 miles. Did this training session represent an equivalent ratio of the distance he ran to the distance he cycled? Explain why or why not.

> **Lesson Summary**
>
> The *value of the ratio* $A : B$ is the quotient $\dfrac{A}{B}$ as long as B is not zero.
>
> If two ratios are equivalent, then their values are the same (when they have values).

Name _____   Date _____

You created a new playlist, and 100 of your friends listened to it and shared if they liked the new playlist or not. Nadhii said the ratio of the number of people who liked the playlist to the number of people who did not like the playlist is 75 : 25. Dylan said that for every three people who liked the playlist, one person did not.

Do Nadhii and Dylan agree? Prove your answer using the values of the ratios.

1. Use the value of the ratio to determine which ratios are equivalent to $9:22$.

   a. $10:23$

   b. $27:66$

   c. $22.5:55$

   d. $4.5:11$

   Answer choices (b), (c), and (d) are equivalent to $9:22$.

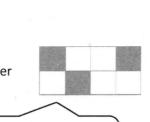

> I can divide 9 by 22 in order to find the value of the ratio, which is $\frac{9}{22}$. To find the value of the ratio for all the answer choices, I need to divide:
>
> $$\frac{10}{23} \neq \frac{9}{22}$$
> $$\frac{27}{66} = \frac{9}{22}$$
> $$\frac{22.5}{55} = \frac{9}{22}$$
> $$\frac{4.5}{11} = \frac{9}{22}$$

2. The ratio of the number of shaded sections to unshaded sections is $3:5$. What is the value of the ratio of the number of shaded sections to the number of unshaded sections?

   $\frac{3}{5}$

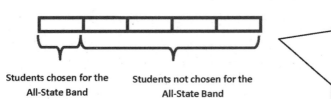

> To find the value of the ratio, I divide 3 by 5. The value of the ratio is $\frac{3}{5}$.

3. The middle school band has 600 members. $\frac{1}{5}$ of the members were chosen for the highly selective All-State Band. What is the value of the ratio of the number of students who were chosen for the All-State Band to the number of students who were not chosen for the All-State Band?

   Students chosen for the All-State Band   Students not chosen for the All-State Band

   The value of the ratio of the number of students who were chosen for the All-State Band to the number of students who were not chosen for the All-State Band is $\frac{1}{4}$.

> In the tape diagram, $\frac{1}{5}$ of the members in the band were chosen for the All-State Band. I can divide 600 by 5, which is 120, so I know 120 students were selected for the All-State Band and 480 were not, because $600 - 120 = 480$. I also know the value of one unit, which is 120, and the value of 4 units, which is 480. The value of the ratio is $\frac{120}{480}$, or $\frac{1}{4}$.

4. Tina is learning to juggle and has set a personal goal of juggling for at least five seconds. She tried 30 times but only accomplished her goal 14 times.

a. Describe and write more than one ratio related to this situation.

The ratio of the number of successful tries to the total number of tries is $14:30$.

The ratio of the number of successful tries to the number of unsuccessful tries is $14:16$.

The ratio of the number of unsuccessful tries to the number of successful tries is $16:14$.

The ratio of the number of unsuccessful tries to the total number of tries is $16:30$.

> There is more than one ratio associated with this problem. I know the total, 30, and the number of times she was successful, 14. I can also determine the number of times she was unsuccessful ($30 - 14 = 16$).

b. For each ratio you created, use the value of the ratio to express one quantity as a fraction of the other quantity.

The number of successful tries is $\frac{14}{30}$, **or** $\frac{7}{15}$, **of the total number of tries.**

The number of successful tries is $\frac{14}{16}$, **or** $\frac{7}{8}$ **the number of unsuccessful tries.**

The number of unsuccessful tries is $\frac{16}{14}$, **or** $\frac{8}{7}$, **the number of successful tries.**

The number of unsuccessful tries is $\frac{16}{30}$, **or** $\frac{8}{15}$, **of the total number of tries.**

c. Create a word problem that a student can solve using one of the ratios and its value.

If Tina tries juggling for at least five seconds 15 times, how many successes would she anticipate having, assuming her ratio of successful tries to unsuccessful tries does not change?

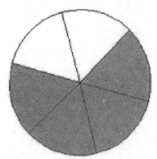

1. The ratio of the number of shaded sections to the number of unshaded sections is 4 to 2. What is the value of the ratio of the number of shaded pieces to the number of unshaded pieces?

2. Use the value of the ratio to determine which ratios are equivalent to $7 : 15$.

    a. $21 : 45$

    b. $14 : 45$

    c. $3 : 5$

    d. $63 : 135$

3. Sean was at batting practice. He swung 25 times but only hit the ball 15 times.

    a. Describe and write more than one ratio related to this situation.

    b. For each ratio you created, use the value of the ratio to express one quantity as a fraction of the other quantity.

    c. Make up a word problem that a student can solve using one of the ratios and its value.

4. Your middle school has 900 students. $\frac{1}{3}$ of students bring their lunch instead of buying lunch at school. What is the value of the ratio of the number of students who do bring their lunch to the number of students who do not?

Example 1

To make paper mache, the art teacher mixes water and flour. For every two cups of water, she needs to mix in three cups of flour to make the paste.

Find equivalent ratios for the ratio relationship 2 cups of water to 3 cups of flour. Represent the equivalent ratios in the table below:

Cups of Water	Cups of Flour

Example 2

Javier has a new job designing websites. He is paid at a rate of $700 for every 3 pages of web content that he builds. Create a ratio table to show the total amount of money Javier has earned in ratio to the number of pages he has built.

Total Pages Built								
Total Money Earned								

Javier is saving up to purchase a used car that costs $4,200. How many web pages will Javier need to build before he can pay for the car?

Exercise 1

Spraying plants with cornmeal juice is a natural way to prevent fungal growth on the plants. It is made by soaking cornmeal in water, using two cups of cornmeal for every nine gallons of water. Complete the ratio table to answer the questions that follow.

Cups of Cornmeal	Gallons of Water

a. How many cups of cornmeal should be added to 45 gallons of water?

b. Paul has only 8 cups of cornmeal. How many gallons of water should he add if he wants to make as much cornmeal juice as he can?

c. What can you say about the values of the ratios in the table?

Exercise 2

James is setting up a fish tank. He is buying a breed of goldfish that typically grows to be 12 inches long. It is recommended that there be 1 gallon of water for every inch of fish length in the tank. What is the recommended ratio of gallons of water per fully-grown goldfish in the tank?

Complete the ratio table to help answer the following questions:

Number of Fish	Gallons of Water

a. What size tank (in gallons) is needed for James to have 5 full-grown goldfish?

b. How many fully-grown goldfish can go in a 40-gallon tank?

c. What can you say about the values of the ratios in the table?

Lesson 9:   Tables of Equivalent Ratios

> **Lesson Summary**
>
> A ratio table is a table of pairs of numbers that form equivalent ratios.

Name _____   Date _____

A father and his young toddler are walking along the sidewalk. For every 3 steps the father takes, the son takes 5 steps just to keep up. What is the ratio of the number of steps the father takes to the number of steps the son takes? Add labels to the columns of the table, and place the ratio into the first row of data. Add equivalent ratios to build a ratio table.

What can you say about the values of the ratios in the table?

Assume the following represents a table of equivalent ratios. Fill in the missing values. Then create a real-world context for the ratios shown in the table.

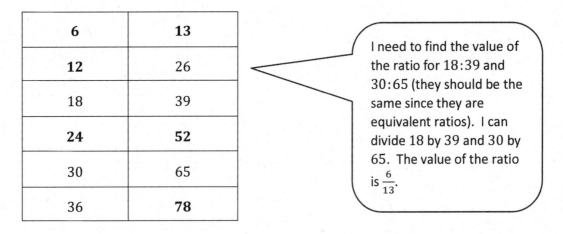

6	13
12	26
18	39
24	52
30	65
36	78

I need to find the value of the ratio for $18:39$ and $30:65$ (they should be the same since they are equivalent ratios). I can divide 18 by 39 and 30 by 65. The value of the ratio is $\frac{6}{13}$.

Sample Answer: Brianna is mixing red and white paint to make a particular shade of pink paint. For every 6 tablespoons of white paint, she mixes 13 tablespoons of red paint. How many tablespoons of red paint would she need for 30 tablespoons of white paint?

Assume each of the following represents a table of equivalent ratios. Fill in the missing values. Then choose one of the tables and create a real-world context for the ratios shown in the table.

1.

	22
12	
16	44
	55
24	66

2.

	14
15	21
25	35
30	

3.

	34
	51
12	
15	85
18	102

Exploratory Challenge

Imagine that you are making a fruit salad. For every quart of blueberries you add, you would like to put in 3 quarts of strawberries. Create three ratio tables that show the amounts of blueberries and strawberries you would use if you needed to make fruit salad for greater numbers of people.

Table 1 should contain amounts where you have added fewer than 10 quarts of blueberries to the salad.

Table 2 should contain amounts of blueberries between and including 10 and 50 quarts.

Table 3 should contain amounts of blueberries greater than or equal to 100 quarts.

Table 1	
Quarts of Blueberries	Quarts of Strawberries

Table 2	
Quarts of Blueberries	Quarts of Strawberries

Table 3	
Quarts of Blueberries	Quarts of Strawberries

a. Describe any patterns you see in the tables. Be specific in your descriptions.

b. How are the amounts of blueberries and strawberries related to each other?

c. How are the values in the Blueberries column related to each other?

d. How are the values in the Strawberries column related to each other?

e. If we know we want to add 7 quarts of blueberries to the fruit salad in Table 1, how can we use the table to help us determine how many strawberries to add?

Lesson 10:   The Structure of Ratio Tables—Additive and Multiplicative

f.   If we know we used 70 quarts of blueberries to make our salad, how can we use a ratio table to find out how many quarts of strawberries were used?

Exercise 1

The following tables were made incorrectly. Find the mistakes that were made, create the correct ratio table, and state the ratio that was used to make the correct ratio table.

a.

Hours	Pay in Dollars
3	24
5	40
7	52
9	72

Hours	Pay in Dollars

Ratio _____

b.

Blue	Yellow
1	5
4	8
7	13
10	16

Blue	Yellow

Ratio _____

Lesson Summary

Ratio tables are constructed in a special way.

Each pair of values in the table will be equivalent to the same ratio.

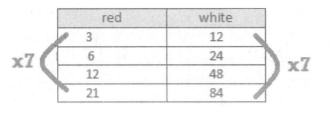

red	white
3	12
6	24
21	84

6 : 24   21 : 84

1 : 4   1 : 4

Repeated addition or multiplication can be used to create a ratio table.

The values in the first column can be multiplied by a constant value to get the values in the second column.

red		white
3	× 4	12
6	× 4	24
12	× 4	48
21	× 4	84

By just adding a certain number to the first entry of a ratio in the first column and adding the same number to the second entry in the second column, the new ratio formed is generally not equivalent to the original ratio. Instead, the numbers added to the entries must be related to the ratio used to make the table. However, if the entries in one column are multiplied by a certain number, multiplying the entries in the other column by the same number creates equivalent ratios.

x7

red	white
3	12
6	24
12	48
21	84

x7

Lesson 10:   The Structure of Ratio Tables—Additive and Multiplicative

Name _____   Date _____

Show more than one way you could use the structure of the table to find the unknown value. Fill in the unknown values.

Number of Weeks	Amount of Money in Account
2	$350
4	$700
6	$1,050
8	
10	

1. Lenard made a table to show how much blue and yellow pain the needs to mix to reach the shade of green he will use to paint the ramps at the skate park. He wants to use the table to make larger and smaller batches of green paint.

Blue	Yellow
10	4
15	6
20	8
25	10

> I see that the value in the first column keeps increasing by 5, and the value in the second column keeps increasing by 2, so the ratio is 5:2. All of the ratios listed in the table are equivalent.

a. What ratio was used to create this table? Support your answer.

The ratio of the amount of blue paint to the amount of yellow paint is $5:2$. $10:4, 15:6, 20:8,$ *and* $25:10$ *are all equivalent to* $5:2$.

b. How are the values in each row related to each other?

In each row, the amount of yellow paint is $\frac{2}{5}$ *the amount of blue paint, or the amount of blue paint is* $\frac{5}{2}$ *the amount of yellow paint.*

c. How are the values in each column related to each other?

The values in the columns are increasing using the ratio. Since the ratio of the amount of blue paint to the amount of yellow paint is $5:2$, *I repeatedly added to form the table.* 5 *was added to the entries in the blue column, and* 2 *was added to the entries in the yellow column.*

2.

a. Create a ratio table for making 2-ingredient banana pancakes with a banana-to-egg ratio of $1:2$. Show how many eggs would be needed to make banana pancakes if you use 14 bananas.

Number of Bananas	Number of Eggs
1	2
2	4
3	6
4	8
14	28

> I need to label the missing title: Number of Eggs. I can complete the table using the relationship: for every 1 banana, I need 2 eggs. I can add 1 repeatedly in the first column and add 2 repeatedly in the second column to determine values in the table. Or, I can multiply the values in the first column by two because the number of eggs is twice the number of bananas.

28 eggs would be needed to make banana pancakes if 14 bananas are used.

b. How is the value of the ratio used to create the table?

The value of the ratio of the number of bananas to the number of eggs is $\frac{1}{2}$. If I know the number of bananas, I can multiply that amount by 2 to get the number of eggs. If I know the number of eggs, I can multiply that amount by $\frac{1}{2}$ (or divide by 2) to get the number of bananas.

Lesson 10:   The Structure of Ratio Tables—Additive and Multiplicative

1.

    a.   Create a ratio table for making lemonade with a lemon juice-to-water ratio of $1:3$. Show how much lemon juice would be needed if you use 36 cups of water to make lemonade.

    b.   How is the value of the ratio used to create the table?

2. Ryan made a table to show how much blue and red paint he mixed to get the shade of purple he will use to paint the room. He wants to use the table to make larger and smaller batches of purple paint.

Blue	Red
12	3
20	5
28	7
36	9

    a.   What ratio was used to create this table? Support your answer.

    b.   How are the values in each row related to each other?

    c.   How are the values in each column related to each other?

Example 1

Create four equivalent ratios (2 by scaling up and 2 by scaling down) using the ratio 30 to 80.

Write a ratio to describe the relationship shown in the table.

Hours	Number of Pizzas Sold
2	16
5	40
6	48
10	80

Exercise 1

The following tables show how many words each person can text in a given amount of time. Compare the rates of texting for each person using the ratio table.

Michaela

Minutes	3	5	7	9
Words	150	250	350	450

Jenna

Minutes	2	4	6	8
Words	90	180	270	360

Maria

Minutes	3	6	9	12
Words	120	240	360	480

Complete the table so that it shows Max has a texting rate of 55 words per minute.

Max

Minutes				
Words				

Exercise 2 : Making Juice (Comparing Juice to Water)

a.   The tables below show the comparison of the amount of water to the amount of juice concentrate (JC) in grape juice made by three different people. Whose juice has the greatest water-to-juice concentrate ratio, and whose juice would taste strongest? Be sure to justify your answer.

Laredo's Juice

Water	JC	Total
12	4	16
15	5	20
21	7	28
45	15	60

Franca's Juice

Water	JC	Total
10	2	12
15	3	18
25	5	30
40	8	48

Milton's Juice

Water	JC	Total
8	2	10
16	4	20
24	6	30
40	10	50

Put the juices in order from the juice containing the most water to the juice containing the least water.

_____   _____   _____

Explain how you used the values in the table to determine the order.

What ratio was used to create each table?

Laredo: _____   Franca: _____

Milton: _____

EUREKA
MATH

Explain how the ratio could help you compare the juices.

b. The next day, each of the three people made juice again, but this time they were making apple juice. Whose juice has the greatest water-to-juice concentrate ratio, and whose juice would taste the strongest? Be sure to justify your answer.

Laredo's Juice		
Water	JC	Total
12	2	14
18	3	21
30	5	35
42	7	49

Franca's Juice		
Water	JC	Total
15	6	21
20	8	28
35	14	49
50	20	70

Milton's Juice		
Water	JC	Total
16	6	22
24	9	33
40	15	55
64	24	88

Put the juices in order from the strongest apple taste to the weakest apple taste.

_____   _____   _____

Explain how you used the values in the table to determine the order.

What ratio was used to create each table?

Laredo: _____   Franca: _____

Milton: _____

Explain how the ratio could help you compare the juices.

How was this problem different than the grape juice questions in part (a)?

c. Max and Sheila are making orange juice. Max has mixed 15 cups of water with 4 cups of juice concentrate. Sheila has made her juice by mixing 8 cups water with 3 cups of juice concentrate. Compare the ratios of juice concentrate to water using ratio tables. State which beverage has a higher juice concentrate-to-water ratio.

d. Victor is making recipes for smoothies. His first recipe calls for 2 cups of strawberries and 7 cups of other ingredients. His second recipe says that 3 cups of strawberries are combined with 9 cups of other ingredients. Which smoothie recipe has more strawberries compared to other ingredients? Use ratio tables to justify your answer.

Lesson Summary

Ratio tables can be used to compare two ratios.

Look for equal amounts in a row or column to compare the second amount associated with it.

3	6	12	30
7	14	28	70

10	25	30	45
16	40	48	72

The values of the tables can also be extended in order to get comparable amounts. Another method would be to compare the values of the ratios by writing the values of the ratios as fractions and then using knowledge of fractions to compare the ratios.

When ratios are given in words, creating a table of equivalent ratios helps in comparing the ratios.

$12 : 35$ compared to $8 : 20$

Quantity 1	12	24	36	48
Quantity 2	35	70	105	140

Quantity 1	8	56
Quantity 2	20	140

Name _____ Date _____

Beekeepers sometimes supplement the diet of honey bees with sugar water to help promote colony growth in the spring and help the bees survive through fall and winter months. The tables below show the amount of water and the amount of sugar used in the Spring and in the Fall.

Spring Sugar Water Mixture	
Sugar (cups)	Water (cups)
6	4
15	10
18	12
27	18

Fall Sugar Water Mixture	
Sugar (cups)	Water (cups)
4	2
10	5
14	7
30	15

Write a sentence that compares the ratios of the number of cups of sugar to the number of cups of water in each table.

Explain how you determined your answer.

1. Jasmine and Juliet were texting.

   a. Use the ratio tables below to determine who texts faster.

   Jasmine

Time (min)	2	5	6	8
Number of Words	56	140	168	224

   > If Jasmine can text 56 words in 2 minutes, I can determine how many words she can text in 1 minute by dividing both numbers by 2.

   Juliet

Time (min)	3	4	7	10
Number of Words	99	132	231	330

   Juliet texts 33 words in 1 minute, which is faster than Jasmine who texts 28 words in 1 minute.

   > If Juliet can text 99 words in 3 minutes, I can determine how many words she texts in 1 minute by dividing both numbers by 3.

   b. Explain the method that you used to determine your answer.

   To determine how many words Jasmine texts in a minute, I divided 56 by 2 since she texted 56 words in 2 minutes. So, Jasmine texts 28 words in 1 minute. For Juliet, I divided 99 by 3 since she texted 99 words in 3 minutes. So, Juliet texts 33 words in 1 minute.

2. Victor is making lemonade. His first recipe calls for 2 cups of water and the juice from 12 lemons. His second recipe says he will need 3 cups of water and the juice from 15 lemons. Use ratio tables to determine which lemonade recipe calls for more lemons compared to water.

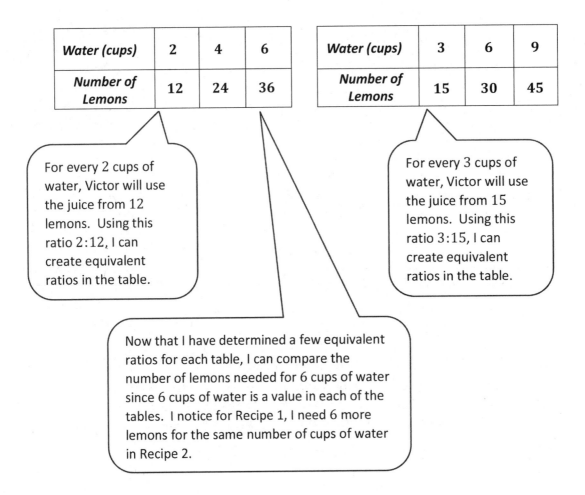

Recipe 1

Water (cups)	2	4	6
Number of Lemons	12	24	36

For every 2 cups of water, Victor will use the juice from 12 lemons. Using this ratio 2:12, I can create equivalent ratios in the table.

Recipe 2

Water (cups)	3	6	9
Number of Lemons	15	30	45

For every 3 cups of water, Victor will use the juice from 15 lemons. Using this ratio 3:15, I can create equivalent ratios in the table.

Now that I have determined a few equivalent ratios for each table, I can compare the number of lemons needed for 6 cups of water since 6 cups of water is a value in each of the tables. I notice for Recipe 1, I need 6 more lemons for the same number of cups of water in Recipe 2.

Recipe 1 uses more lemons compared to water. When comparing 6 cups of water, there were more lemons used in Recipe 1 than in Recipe 2.

1. Sarah and Eva were swimming.

   a. Use the ratio tables below to determine who the faster swimmer is.

   Sarah

Time (min)	3	5	12	17
Distance (meters)	75	125	300	425

   Eva

Time (min)	2	7	10	20
Distance (meters)	52	182	260	520

   b. Explain the method that you used to determine your answer.

   a. A 120 lb. person would weigh about 20 lb. on the earth's moon. A 150 lb. person would weigh 28 lb. on Io, a moon of Jupiter. Use ratio tables to determine which moon would make a person weigh the most.

Exercise 2

The amount of sugary beverages Americans consume is a leading health concern. For a given brand of cola, a 12 oz. serving of cola contains about 40 g of sugar. Complete the ratio table, using the given ratio to find equivalent ratios.

Cola (ounces)		12	
Sugar (grams)		40	

Exercise 3

A 1 L bottle of cola contains approximately 34 fluid ounces. How many grams of sugar would be in a 1 L bottle of the cola? Explain and show how to arrive at the solution.

Exercise 4

A school cafeteria has a restriction on the amount of sugary drinks available to students. Drinks may not have more than 25 g of sugar. Based on this restriction, what is the largest size cola (in ounces) the cafeteria can offer to students?

Exercise 5

Shontelle solves three math problems in four minutes.

a. Use this information to complete the table below.

Number of Questions	3	6	9	12	15	18	21	24	27	30
Number of Minutes										

b. Shontelle has soccer practice on Thursday evening. She has a half hour before practice to work on her math homework and to talk to her friends. She has 20 math skill-work questions for homework, and she wants to complete them before talking with her friends. How many minutes will Shontelle have left after completing her math homework to talk to her friends?

Use a double number line diagram to support your answer, and show all work.

Lesson Summary

A *double number line* is a representation of a ratio relationship using a pair of parallel number lines. One number line is drawn above the other so that the zeros of each number line are aligned directly with each other. Each ratio in a ratio relationship is represented on the double number line by always plotting the first entry of the ratio on one of the number lines and plotting the second entry on the other number line so that the second entry is aligned with the first entry.

Double Number Line Reproducible

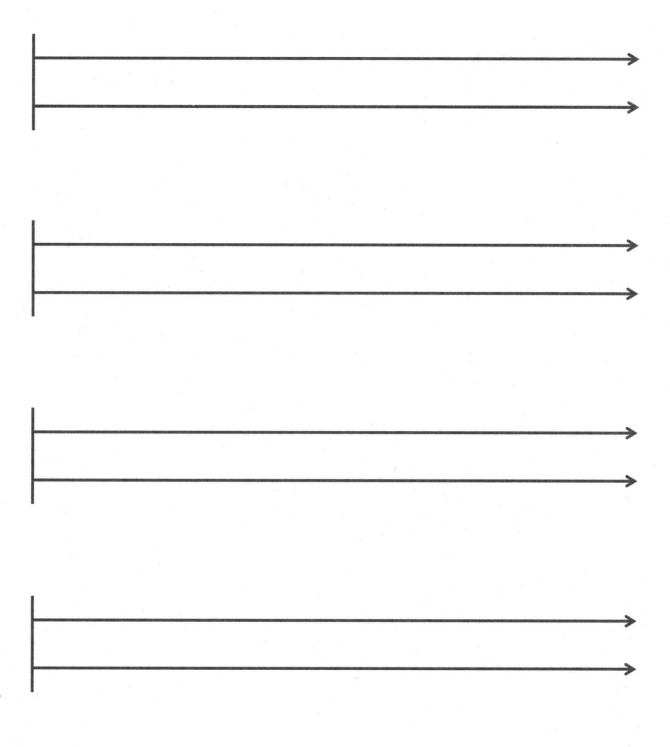

Name _____   Date _____

Kyra is participating in a fundraiser walk-a-thon. She walks 2 miles in 30 minutes. If she continues to walk at the same rate, determine how many minutes it will take her to walk 7 miles. Use a double number line diagram to support your answer.

1. David earns $6 an hour for helping with yard work. He wants to buy a new video game that costs $27. How many hours must he help in the yard to earn $27 to buy the game? Use a double number line diagram to support your answer.

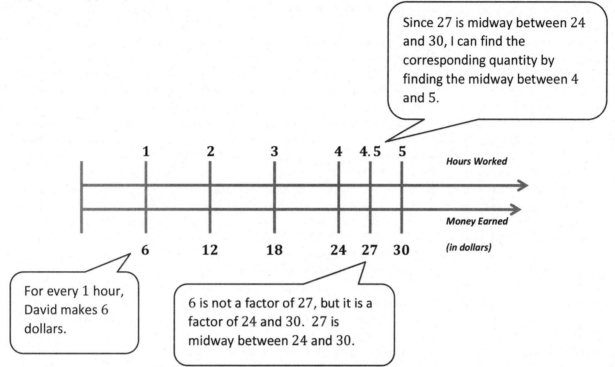

Since 27 is midway between 24 and 30, I can find the corresponding quantity by finding the midway between 4 and 5.

For every 1 hour, David makes 6 dollars.

6 is not a factor of 27, but it is a factor of 24 and 30. 27 is midway between 24 and 30.

David will earn $27 after working for 4.5 hours.

2. During migration, a duck flies at a constant rate for 11 hours, during which time he travels 550 miles. The duck must travel another 250 miles in order to reach his destination. If the duck maintains the same constant speed, how long will it take him to complete the remaining 250 miles? Include a table or diagram to support your answer.

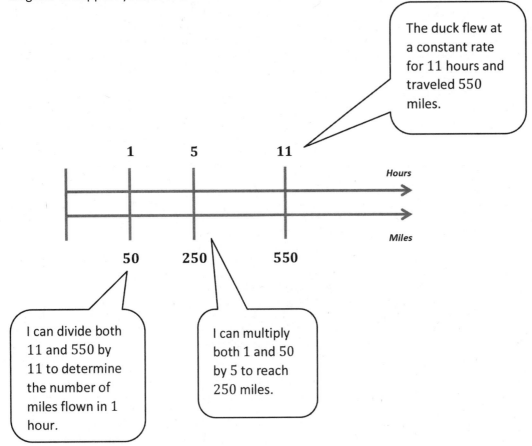

It will take the duck 5 *hours to travel the remaining* 250 *miles*.

1. While shopping, Kyla found a dress that she would like to purchase, but it costs $52.25 more than she has. Kyla charges $5.50 an hour for babysitting. She wants to figure out how many hours she must babysit to earn $52.25 to buy the dress. Use a double number line to support your answer.

2. Frank has been driving at a constant speed for 3 hours, during which time he traveled 195 miles. Frank would like to know how long it will take him to complete the remaining 455 miles, assuming he maintains the same constant speed. Help Frank determine how long the remainder of the trip will take. Include a table or diagram to support your answer.

Exercise 1

Jorge is mixing a special shade of orange paint. He mixed 1 gallon of red paint with 3 gallons of yellow paint.

Based on this ratio, which of the following statements are true?

- $\frac{3}{4}$ of a 4-gallon mix would be yellow paint.

- Every 1 gallon of yellow paint requires $\frac{1}{3}$ gallon of red paint.

- Every 1 gallon of red paint requires 3 gallons of yellow paint.

- There is 1 gallon of red paint in a 4-gallon mix of orange paint.

- There are 2 gallons of yellow paint in an 8-gallon mix of orange paint.

Use the space below to determine if each statement is true or false.

Exercise 2

Based on the information on red and yellow paint given in Exercise 1, complete the table below.

Red Paint (R)	Yellow Paint (Y)	Relationship
	3	$3 = 1 \times 3$
2		
	9	$9 = 3 \times 3$
	12	
5		

Exercise 3

a. Jorge now plans to mix red paint and blue paint to create purple paint. The color of purple he has decided to make combines red paint and blue paint in the ratio $4 : 1$. If Jorge can only purchase paint in one gallon containers, construct a ratio table for all possible combinations for red and blue paint that will give Jorge no more than 25 gallons of purple paint.

Blue (B)	Red (R)	Relationship

Write an equation that will let Jorge calculate the amount of red paint he will need for any given amount of blue paint.

Write an equation that will let Jorge calculate the amount of blue paint he will need for any given amount of red paint.

If Jorge has 24 gallons of red paint, how much blue paint will he have to use to create the desired color of purple?

If Jorge has 24 gallons of blue paint, how much red paint will he have to use to create the desired color of purple?

b. Using the same relationship of red to blue from above, create a table that models the relationship of the three colors blue, red, and purple (total) paint. Let B represent the number of gallons of blue paint, let R represent the number of gallons of red paint, and let T represent the total number of gallons of (purple) paint. Then write an equation that models the relationship between the blue paint and the total paint, and answer the questions.

Blue (B)	Red (R)	Total Paint (T)

Equation:

Value of the ratio of total paint to blue paint:

How is the value of the ratio related to the equation?

Excercise 4

During a particular U.S. Air Force training exercise, the ratio of the number of men to the number of women was 6 : 1. Use the ratio table provided below to create at least two equations that model the relationship between the number of men and the number of women participating in this training exercise.

Women (W)	Men (M)

Equations :

If 200 women participated in the training exercise, use one of your equations to calculate the number of men who participated.

Exercise 5

Malia is on a road trip. During the first five minutes of Malia's trip, she sees 18 cars and 6 trucks. Assuming this ratio of cars to trucks remains constant over the duration of the trip, complete the ratio table using this comparison. Let T represent the number of trucks she sees, and let C represent the number of cars she sees.

Trucks (T)	Cars (C)
1	
3	
	18
12	
	60

What is the value of the ratio of the number of cars to the number of trucks?

What equation would model the relationship between cars and trucks?

At the end of the trip, Malia had counted 1,254 trucks. How many cars did she see?

Excercise 6

Kevin is training to run a half-marathon. His training program recommends that he run for 5 minutes and walk for 1 minute. Let R represent the number of minutes running, and let W represent the number of minutes walking.

Minutes Running (R)		10	20		50
Minutes Walking (W)	1	2		8	

What is the value of the ratio of the number of minutes walking to the number of minutes running?

What equation could you use to calculate the minutes spent walking if you know the minutes spent running?

Lesson Summary

The value of a ratio can be determined using a ratio table. This value can be used to write an equation that also represents the ratio.

Example:

1	4
2	8
3	12
4	16

The multiplication table can be a valuable resource to use in seeing ratios. Different rows can be used to find equivalent ratios.

Name _____   Date _____

A carpenter uses four nails to install each shelf. Complete the table to represent the relationship between the number of nails (N) and the number of shelves (S). Write the ratio that describes the number of nails per number of shelves. Write as many different equations as you can that describe the relationship between the two quantities.

Shelves (S)	Nails (N)
1	4
2	
	12
	16
5	

A pie recipe calls for 2 teaspoons of cinnamon and 3 teaspoons of nutmeg.

Make a table showing the comparison of the number of teaspoons of cinnamon and the number of teaspoons of nutmeg.

Number of Teaspoons of Cinnamon (C)	Number of Teaspoons of Nutmeg (N)
2	3
4	6
6	9
8	12
10	15

I know the ratio of the number of teaspoons of cinnamon to the number of teaspoons of nutmeg is $2 : 3$ because that information is given in the problem. I will write this ratio in the first row of the table and then determine equivalent ratios.

1. Write the value of the ratio of the number of teaspoons of cinnamon to the number of teaspoons of nutmeg.

$\dfrac{2}{3}$

Anytime I see a ratio relationship, I pay close attention to the order. In this problem, I'm comparing the number of teaspoons of cinnamon to the number of teaspoons of nutmeg. So, I look at the first row in my table. The numerator is the number of teaspoons of cinnamon, which is 2, and the denominator is the number of teaspoons of nutmeg, which is 3.

2. Write an equation that shows the relationship of the number of teaspoons of cinnamon to the number of teaspoons of nutmeg.

$N = \frac{3}{2}C$ or $C = \frac{2}{3}N$

> To write an equation, I have to pay close attention to the value of the ratio for teaspoons of nutmeg to teaspoons of cinnamon, which is $\frac{3}{2}$. Now, I can write the equation $N = \frac{3}{2}C$.

3. Explain how the value of the ratio of the number of teaspoons of nutmeg to the number of teaspoons of cinnamon can be seen in the table.

The values in the first row show the values in the ratio. The ratio of the number of teaspoons of nutmeg to the number of teaspoons of cinnamon is $3:2$. The value of the ratio is $\frac{3}{2}$.

4. Explain how the value of the ratio of the number of teaspoons of nutmeg to the number of teaspoons of cinnamon can be seen in an equation.

The number of teaspoons of nutmeg is represented as N in the equation. The number of teaspoons of cinnamon is represented as C. The value of the ratio is represented because the number of teaspoons of nutmeg is $\frac{3}{2}$ times as much as the number of teaspoons of cinnamon, $N = \frac{3}{2}C$.

5. Using the same recipe, compare the number of teaspoons of cinnamon to the number of total teaspoons of spices used in the recipe.

Make a table showing the comparison of the number of teaspoons of cinnamon to the number of total teaspoons of spices.

Number of Teaspoons of Cinnamon (C)	Number of Total Teaspoons of Spices (T)
2	5
4	10
6	15
8	20
10	25

> To get the total, I will look at the table I made on the previous page. I see that for every 2 teaspoons of cinnamon, I will need 3 teaspoons of nutmeg, so for 2 teaspoons of cinnamon, there are 5 total teaspoons of spices since $2 + 3 = 5$.

Lesson 13:   From Ratio Tables to Equations Using the Value of a Ratio

6. Write the value of the ratio of the number of total teaspoons of spices to the number of teaspoons of cinnamon.

$\frac{5}{2}$

> I will look at the first row. There are 5 total teaspoons of spices and 2 teaspoons of cinnamon. Now I can write the value of the ratio.

7. Write an equation that shows the relationship of the number of total teaspoons of spices to the number of teaspoons of cinnamon.

$T = \frac{5}{2}C$

> To write this equation, I will use the value of the ratio that I determined from Problem 5, which is $\frac{5}{2}$.

A cookie recipe calls for 1 cup of white sugar and 3 cups of brown sugar.

Make a table showing the comparison of the amount of white sugar to the amount of brown sugar.

White Sugar (W)	Brown Sugar (B)

1. Write the value of the ratio of the amount of white sugar to the amount of brown sugar.

2. Write an equation that shows the relationship of the amount of white sugar to the amount of brown sugar.

3. Explain how the value of the ratio can be seen in the table.

4. Explain how the value of the ratio can be seen in the equation.

Using the same recipe, compare the amount of white sugar to the amount of total sugars used in the recipe.

Make a table showing the comparison of the amount of white sugar to the amount of total sugar.

White Sugar (W)	Total Sugar (T)

5. Write the value of the ratio of the amount of total sugar to the amount of white sugar.

6. Write an equation that shows the relationship of total sugar to white sugar.

Kelli is traveling by train with her soccer team from Yonkers, NY to Morgantown, WV for a tournament. The distance between Yonkers and Morgantown is 400 miles. The total trip will take 8 hours. The train schedule is provided below:

Leaving Yonkers, New York	
Destination	Distance
Allentown, PA	100 miles
Carlisle, PA	200 miles
Berkeley Springs, WV	300 miles
Morgantown, WV	400 miles

Leaving Morgantown, WV	
Destination	Distance
Berkeley Springs, WV	100 miles
Carlisle, PA	200 miles
Allentown, PA	300 miles
Yonkers, NY	400 miles

Exercises

1. Create a table to show the time it will take Kelli and her team to travel from Yonkers to each town listed in the schedule assuming that the ratio of the amount of time traveled to the distance traveled is the same for each city. Then, extend the table to include the cumulative time it will take to reach each destination on the ride home.

Hours	Miles

2. Create a double number line diagram to show the time it will take Kelli and her team to travel from Yonkers to each town listed in the schedule. Then, extend the double number line diagram to include the cumulative time it will take to reach each destination on the ride home. Represent the ratio of the distance traveled on the round trip to the amount of time taken with an equation.

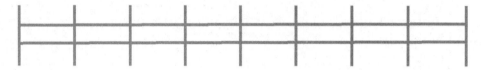

Using the information from the double number line diagram, how many miles would be traveled in one hour?

_____

How do you know?

_____

_____

Example 1

Dinner service starts once the train is 250 miles away from Yonkers. What is the minimum time the players will have to wait before they can have their meal?

Hours	Miles	Ordered Pairs

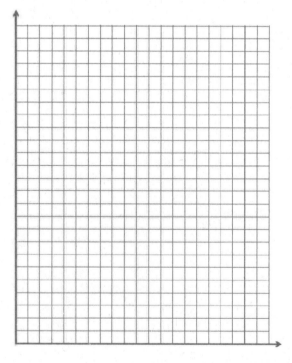

      Lesson 14:   From Ratio Tables, Equations, and Double Number Line Diagrams to Plots on the Coordinate Plane

Lesson Summary

A ratio table, equation, or double number line diagram can be used to create ordered pairs. These ordered pairs can then be graphed on a coordinate plane as a representation of the ratio.

Example:

Equation: $y = 3x$

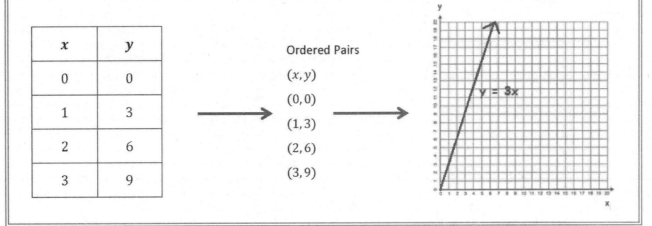

x	y
0	0
1	3
2	6
3	9

Ordered Pairs

(x, y)

$(0, 0)$

$(1, 3)$

$(2, 6)$

$(3, 9)$

Name _____   Date _____

Dominic works on the weekends and on vacations from school mowing lawns in his neighborhood. For every lawn he mows, he charges $12. Complete the table. Then determine ordered pairs, and create a labeled graph.

Lawns	Charge (in dollars)	Ordered Pairs
2		
4		
6		
8		
10		

1.   How many lawns will Dominic need to mow in order to make $240?

2.   How much money will Dominic make if he mows 9 lawns?

1. Write a story context that would be represented by the ratio 1 : 7.

   Answers will vary. Example: For every hour Sami rakes leaves, she earns $7.

   > I can think of a situation that compares 1 of one quantity to 7 of another quantity. For every 1 hour she rakes leaves, Sami earns $7.

   Complete a table of values and graph.

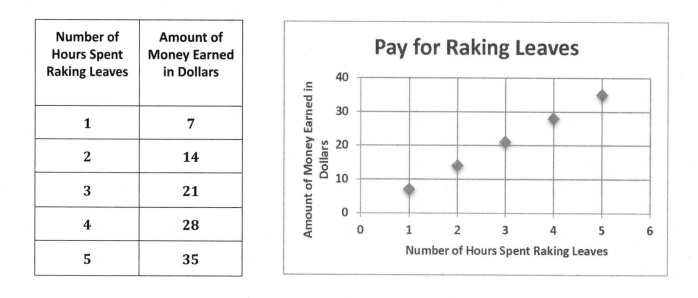

Number of Hours Spent Raking Leaves	Amount of Money Earned in Dollars
1	7
2	14
3	21
4	28
5	35

Pay for Raking Leaves

EUREKA MATH

2. Complete the table of values to find the following:

Find the number of cups of strawberries needed if for every jar of jam Sarah makes, she has to use 5 cups of strawberries.

Number of Jars of Jam	Number of Cups of Strawberries
1	5
2	10
3	15
4	20
5	25

> I can start with the ratio I know from the problem. For every 1 jar of jam, Sarah uses 5 cups of strawberries, so the ratio is 1:5, and I will write this ratio in the first row of my table. I can use this information to determine equivalent ratios.

Use a graph to represent the relationship.

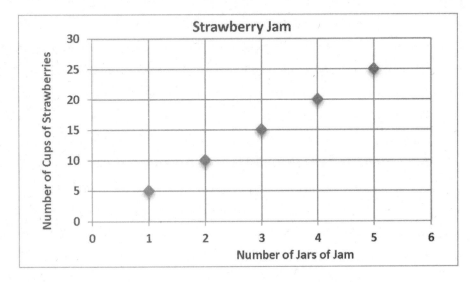

Lesson 14: From Ratio Tables, Equations, and Double Number Line
Diagrams to Plots on the Coordinate Plane

© 2019 Great Minds®. eureka-math.org

EUREKA
MATH®

Create a double number line diagram to show the relationship.

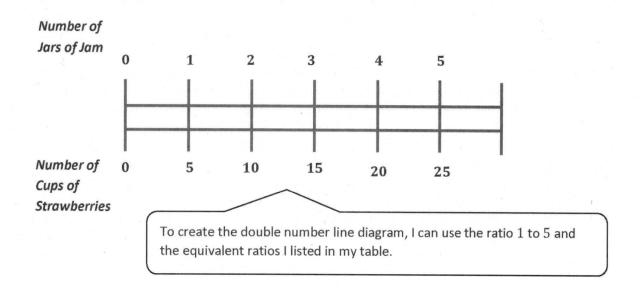

Number of Jars of Jam

0   1   2   3   4   5

Number of Cups of Strawberries

0   5   10   15   20   25

To create the double number line diagram, I can use the ratio 1 to 5 and the equivalent ratios I listed in my table.

1. Complete the table of values to find the following:

   Find the number of cups of sugar needed if for each pie Karrie makes, she has to use 3 cups of sugar.

Pies	Cups of Sugar
1	
2	
3	
4	
5	
6	

   Use a graph to represent the relationship.

   Create a double number line diagram to show the relationship.

2. Write a story context that would be represented by the ratio $1 : 4$.

    Complete a table of values for this equation and graph.

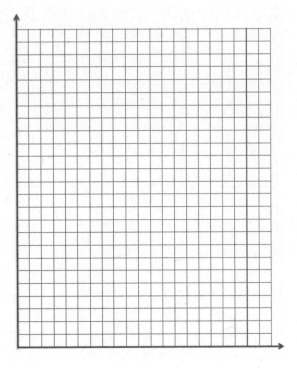

Lesson 14:   From Ratio Tables, Equations, and Double Number Line
                            Diagrams to Plots on the Coordinate Plane

Exploratory Challenge

At the end of this morning's news segment, the local television station highlighted area pets that need to be adopted. The station posted a specific website on the screen for viewers to find more information on the pets shown and the adoption process. The station producer checked the website two hours after the end of the broadcast and saw that the website had 24 views. One hour after that, the website had 36 views.

Exercise 1

Create a table to determine how many views the website probably had one hour after the end of the broadcast based on how many views it had two and three hours after the end of the broadcast. Using this relationship, predict how many views the website will have 4, 5, and 6 hours after the end of the broadcast.

Exercise 2

What is the constant number, c, that makes these ratios equivalent?

Using an equation, represent the relationship between the number of views, v, the website received and the number of hours, h, after this morning's news broadcast.

Exercise 3

Use the table created in Exercise 1 to identify sets of ordered pairs that can be graphed.

Exercise 4

Use the ordered pairs you created to depict the relationship between hours and number of views on a coordinate plane. Label your axes and create a title for the graph. Do the points you plotted lie on a line?

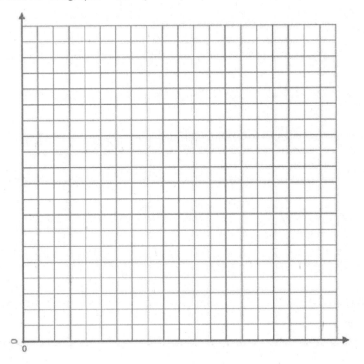

Exercise 5

Predict how many views the website will have after twelve hours. Use at least two representations (e.g., tape diagram, table, double number line diagram) to justify your answer.

Exercise 6

Also on the news broadcast, a chef from a local Italian restaurant demonstrated how he makes fresh pasta daily for his restaurant. The recipe for his pasta is below:

3 eggs, beaten

1 teaspoon salt

2 cups all-purpose flour

2 tablespoons water

2 tablespoons vegetable oil

Determine the ratio of the number of tablespoons of water to the number of eggs.

Provided the information in the table below, complete the table to determine ordered pairs. Use the ordered pairs to graph the relationship of the number of tablespoons of water to the number of eggs.

Tablespoons of water	Number of Eggs
2	
4	
6	
8	
10	
12	

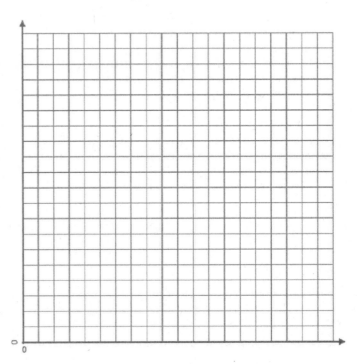

What would you have to do to the graph in order to find how many eggs would be needed if the recipe was larger and called for 16 tablespoons of water?

Demonstrate on your graph.

How many eggs would be needed if the recipe called for 16 tablespoons of water?

Lesson 15:   A Synthesis of Representations of Equivalent Ratio Collections

EUREKA MATH

Exercise 7

Determine how many tablespoons of water will be needed if the chef is making a large batch of pasta and the recipe increases to 36 eggs. Support your reasoning using at least one diagram you find applies best to the situation, and explain why that tool is the best to use.

Lesson Summary

There are several ways to represent the same collection of equivalent ratios. These include ratio tables, tape diagrams, double number line diagrams, equations, and graphs on coordinate planes.

Graph Reproducible

Name _____   Date _____

Jen and Nikki are making bracelets to sell at the local market. They determined that each bracelet would have eight beads and two charms.

Complete the table below to show the ratio of the number of charms to the number of beads.

Charms	2	4	6	8	10
Beads	8				

Create ordered pairs from the table, and plot the pairs on the graph below. Label the axes of the graph, and provide a title.

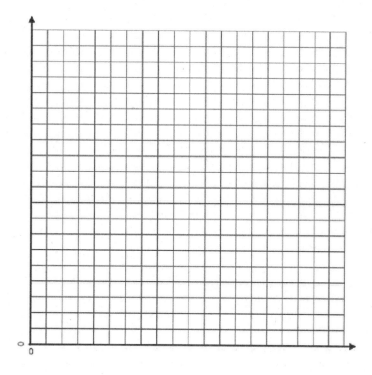

1. When the video of Tillman the Skateboarding Bulldog was first posted, it had 300 views after 4 hours. Create a table to show how many views the video would have after the first, second, and third hours after posting, if the video receives views at the same rate. How many views would the video receive after 5 hours?

Number of Hours	Number of views
1	75
2	150
3	225
4	300
5	375

> First, I can record the information I know in the table. I know there were 300 views after 4 hours. I will determine how many views there were after 1 hour by dividing 300 by 4, which is 75, so there were 75 views after 1 hour. Knowing there were 75 views in 1 hour will allow me to figure out how many views there were after 2 hours, 3 hours, and 5 hours by multiplying the number of hours by 75.

After five hours, the video would have 375 views.

2. Write an equation that represents the relationship from Problem 1. Do you see any connections between the equation you wrote and the ratio of the number of views to the number of hours?

$v = 75h$ The constant in the equation, 75, is the number of views after 1 hour.

> To write the equation, I determine variables to represent the number of hours and the number of views. I can choose v to represent the number of views and h to represent the number of hours. Since the number of views is dependent on how many hours passed since the video was posted, the number of views is the dependent variable and the number of hours is the independent variable. To find out the number of views, I will multiply the number of hours by 75, the constant.

© 2019 Great Minds®. eureka-math.org

3. Use the table in Problem 1 to make a list of ordered pairs that you could plot on a coordinate plane.

   (1, 75), (2, 150), (3, 225), (4, 300), (5, 375)

4. Graph the ordered pairs on a coordinate plane. Label your axes, and create a title for the graph.

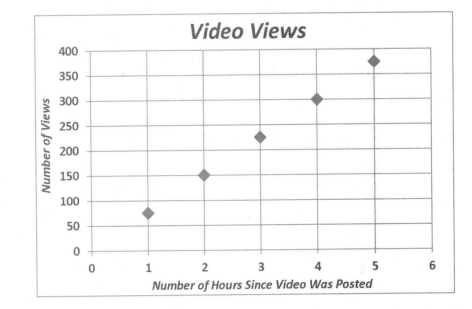

5. Use multiple tools to predict how many views the website would have after 15 hours.

   Answers may vary but could include all representations from the module. The correct answer is **1,125** *views.*

   - *If the equation v = 75h is used, multiply 75 by the number of hours, which is 15. So,*
     $$75 \times 15 = 1,125.$$
   - *To determine the answer using the table, extend the table to show the number of views after 15 hours.*
   - *To use the graph, extend the x-axis to show 15 hours, and then plot the points for 7 through 15 hours since those values are not currently on the graph.*
   - *In a tape diagram, 1 unit has a value of 75 since there were 75 views after the video was posted for 1 hour, so 15 units has a value of 1,125 because 75 × 15 = 1,125.*

Lesson 15:   A Synthesis of Representations of Equivalent Ratio Collections

1. The producer of the news station posted an article about the high school's football championship ceremony on a new website. The website had 500 views after four hours. Create a table to show how many views the website would have had after the first, second, and third hours after posting, if the website receives views at the same rate. How many views would the website receive after 5 hours?

2. Write an equation that represents the relationship from Problem 1. Do you see any connections between the equations you wrote and the ratio of the number of views to the number of hours?

3. Use the table in Problem 1 to make a list of ordered pairs that you could plot on a coordinate plane.

4. Graph the ordered pairs on a coordinate plane. Label your axes and create a title for the graph.

5. Use multiple tools to predict how many views the website would have after 12 hours.

Ratios can be transformed to rates and unit rates.

Example: Introduction to Rates and Unit Rates

Diet cola was on sale last week; it cost $10 for every 4 packs of diet cola.

    a.   How much do 2 packs of diet cola cost?

    b.   How much does 1 pack of diet cola cost?

Exploratory Challenge

    a.   Teagan went to Gamer Realm to buy new video games. Gamer Realm was having a sale: $65 for 4 video games. He bought 3 games for himself and one game for his friend, Diego, but Teagan does not know how much Diego owes him for the one game. What is the unit price of the video games? What is the rate unit?

b.   Four football fans took turns driving the distance from New York to Oklahoma to see a big game. Each driver set the cruise control during his or her portion of the trip, enabling him or her to travel at a constant speed. The group changed drivers each time they stopped for gas and recorded their driving times and distances in the table below.

Fan	Distance (miles)	Time (hours)
Andre	208	4
Matteo	456	8
Janaye	300	6
Greyson	265	5

Use the given data to answer the following questions.

i.   What two quantities are being compared?

ii.   What is the ratio of the two quantities for Andre's portion of the trip? What is the associated rate?

Andre's Ratio: _____   Andre's Rate: _____

iii.   Answer the same two questions in part (ii) for the other three drivers.

Matteo's Ratio: _____   Matteo's Rate: _____

Janaye's Ratio: _____   Janaye's Rate: _____

Greyson's Ratio: _____   Greyson's Rate: _____

iv.   For each driver in parts (ii) and (iii), circle the unit rate and put a box around the rate unit.

c.   A publishing company is looking for new employees to type novels that will soon be published. The publishing company wants to find someone who can type at least 45 words per minute. Dominique discovered she can type at a constant rate of 704 words in 16 minutes. Does Dominique type at a fast enough rate to qualify for the job? Explain why or why not.

Lesson Summary

A *rate* is a quantity that describes a ratio relationship between two types of quantities.

For example, 15 miles/hour is a rate that describes a ratio relationship between hours and miles: If an object is traveling at a constant 15 miles/hour, then after 1 hour it has gone 15 miles, after 2 hours it has gone 30 miles, after 3 hours it has gone 45 miles, and so on.

When a rate is written as a measurement, the *unit rate* is the measure (i.e., the numerical part of the measurement). For example, when the rate of speed of an object is written as the measurement 15 miles/hour, the number 15 is the unit rate. The *unit of measurement* is miles/hour, which is read as "miles per hour."

Name _____   Date _____

Angela enjoys swimming and often swims at a steady pace to burn calories. At this pace, Angela can swim 1,700 meters in 40 minutes.

    a.   What is Angela's unit rate?

    b.   What is the rate unit?

1. The Canter family is downsizing and saving money when they grocery shop. In order to do that, they need to know how to find better prices. At the grocery store downtown, grapes cost $2.55 for 2 lb., and at the farmer's market, grapes cost $ 3.55 for 3 lb.

   a. What is the unit price of grapes at each store? If necessary, round to the nearest penny.

Grocery Store

Number of Pounds of Grapes	1	2
Cost (in dollars)	1.28	2.55

Farmer's Markes

Number of Pounds of Grapes	1	3
Cost (in dollars)	1.18	3.55

I know the price of two pounds. To find the price of one pound, I need to divide the cost by two. $2.55 divided by 2 is $1.275. I need to round to the nearest penny, so the price of one pound is $1.28.

The unit price for the grapes at the grocery store is $1.28.

The unit price for the grapes at the farmer's market is $1.18.

   b. If the Canter family wants to save money, where should they purchase grapes?

   The Canter family should purchase the grapes from the farmer's market. Their unit price is lower, so they pay less money per pound than if they would purchase grapes from the grocery store downtown.

2. Oranges are on sale at the grocery store downtown and at the farmer's market. At the grocery store, a 4 lb. bag of oranges costs $4.99, and at the farmer's market, the price for a 10 lb. bag of oranges is $11.99. Which store offers the best deal on oranges? How do you know? How much better is the deal?

Grocery Store

Number of Pounds of Oranges	1	4
Cost (in dollars)	1.25	4.99

> Now I see that I can divide the cost by the number of pounds to determine the unit price, or the cost of one pound.

Father's Market

Number of Pounds of Oranges	1	10
Cost (in dollars)	1.20	11.99

The unit price for the oranges at the grocery store is $1.25. The unit price for the oranges at the farmer's market is $1.20. The farmer's market offers a better deal on the oranges. Their price is $0.05 cheaper per pound than the price per pound at the grocery store.

> To determine how much better the deal is, I need to subtract the smaller unit price from the larger unit price to find the difference. $1.25 − $1.20 = $0.05. The farmer's market's unit price is five cents cheaper than the grocery store's unit price.

The Scott family is trying to save as much money as possible. One way to cut back on the money they spend is by finding deals while grocery shopping; however, the Scott family needs help determining which stores have the better deals.

1. At Grocery Mart, strawberries cost $2.99 for 2 lb., and at Baldwin Hills Market strawberries are $3.99 for 3 lb.

    a. What is the unit price of strawberries at each grocery store? If necessary, round to the nearest penny.

    b. If the Scott family wanted to save money, where should they go to buy strawberries? Why?

2. Potatoes are on sale at both Grocery Mart and Baldwin Hills Market. At Grocery Mart, a 5 lb. bag of potatoes cost $2.85, and at Baldwin Hills Market a 7 lb. bag of potatoes costs $4.20. Which store offers the best deal on potatoes? How do you know? How much better is the deal?

Given a rate, you can calculate the unit rate and associated ratios. Recognize that all ratios associated with a given rate are equivalent because they have the same value.

Example 1

Write each ratio as a rate.

a. The ratio of miles to the number of hours is 434 to 7.

b. The ratio of the number of laps to the number of minutes is 5 to 4.

Example 2

a. Complete the model below using the ratio from Example 1, part (b).

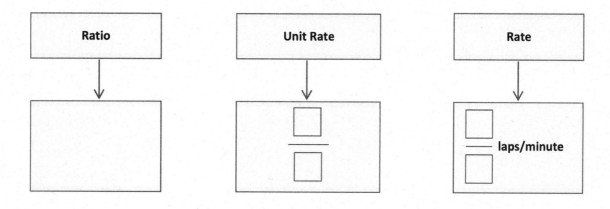

b. Complete the model below now using the rate listed below.

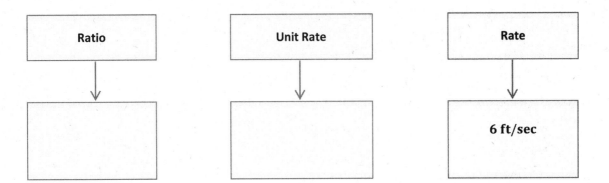

Examples 3–6

3. Dave can clean pools at a constant rate of $\frac{3}{5}$ pools/hour.

a. What is the ratio of the number of pools to the number of hours?

b. How many pools can Dave clean in 10 hours?

c. How long does it take Dave to clean 15 pools?

4. Emeline can type at a constant rate of $\frac{1}{4}$ pages/minute.

    a. What is the ratio of the number of pages to the number of minutes?

    b. Emeline has to type a 5-page article but only has 18 minutes until she reaches the deadline. Does Emeline have enough time to type the article? Why or why not?

    c. Emeline has to type a 7-page article. How much time will it take her?

5. Xavier can swim at a constant speed of $\frac{5}{3}$ meters/second.

    a. What is the ratio of the number of meters to the number of seconds?

    b. Xavier is trying to qualify for the National Swim Meet. To qualify, he must complete a 100-meter race in 55 seconds. Will Xavier be able to qualify? Why or why not?

    c. Xavier is also attempting to qualify for the same meet in the 200-meter event. To qualify, Xavier would have to complete the race in 130 seconds. Will Xavier be able to qualify in this race? Why or why not?

6. The corner store sells apples at a rate of 1.25 dollars per apple.

    a.   What is the ratio of the amount in dollars to the number of apples?

    b.   Akia is only able to spend $10 on apples. How many apples can she buy?

    c.   Christian has $6 in his wallet and wants to spend it on apples. How many apples can Christian buy?

Lesson Summary

A rate of $\frac{2}{3}$ gal/min corresponds to the unit rate of $\frac{2}{3}$ and also corresponds to the ratio $2:3$.

All ratios associated with a given rate are equivalent because they have the same value.

Name _____   Date _____

Tiffany is filling her daughter's pool with water from a hose. She can fill the pool at a rate of $\frac{1}{10}$ gallons/second.

Create at least three equivalent ratios that are associated with the rate. Use a double number line to show your work.

Examples

1. An express train travels at a cruising rate of $150 \frac{\text{miles}}{\text{hour}}$. If the train travels at this average speed for 6 hours, how far does the train travel while at this speed?

Number of Miles	150	300	450	600	750	900
Number of Hours	1	2	3	4	5	6

> I know the ratio of the number of miles to the number of hours is $150:1$. I can create equivalent ratios in a ratio table to determine how many miles the train will travel in 6 hours. $1 \times 6 = 6$, so $150 \times 6 = 900$.

*The train will travel **900** miles in 6 hours traveling at this average cruising speed.*

2. The average amount of rainfall in Baltimore, Maryland in the month of April is $\frac{1}{10} \frac{\text{inch}}{\text{day}}$. Using this rate, how many inches of rain does Baltimore receive on average for the month of April?

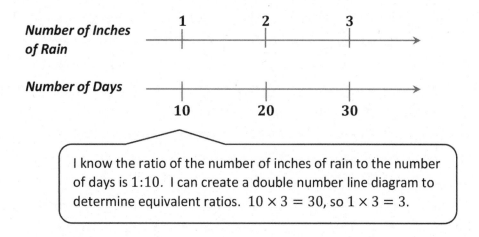

Number of Inches of Rain — 1, 2, 3

Number of Days — 10, 20, 30

> I know the ratio of the number of inches of rain to the number of days is $1:10$. I can create a double number line diagram to determine equivalent ratios. $10 \times 3 = 30$, so $1 \times 3 = 3$.

*At this rate, Baltimore receives **3** inches of rain in the month of April.*

1. Once a commercial plane reaches the desired altitude, the pilot often travels at a cruising speed. On average, the cruising speed is 570 miles/hour. If a plane travels at this cruising speed for 7 hours, how far does the plane travel while cruising at this speed?

2. Denver, Colorado often experiences snowstorms resulting in multiple inches of accumulated snow. During the last snow storm, the snow accumulated at $\frac{4}{5}$ inch/hour. If the snow continues at this rate for 10 hours, how much snow will accumulate?

Mathematical Modeling Exercises

1. At Fun Burger, the Burger Master can make hamburgers at a rate of 4 burgers/minute. In order to address the heavy volume of customers, he needs to continue at this rate for 30 minutes. If he continues to make hamburgers at this pace, how many hamburgers will the Burger Master make in 30 minutes?

2. Chandra is an editor at the New York Gazette. Her job is to read each article before it is printed in the newspaper. If Chandra can read 10 words/second, how many words can she read in 60 seconds?

Exercises

Use the table below to write down your work and answers for the stations.

1.
2.
3.
4.
5.
6.

Lesson 18:   Finding a Rate by Dividing Two Quantities

Lesson Summary

We can convert measurement units using rates. The information can be used to further interpret the problem. Here is an example:

$$\left(5\,\frac{\text{gal}}{\text{min}}\right) \cdot (10\ \text{min}) = \frac{5\ \text{gal}}{1\ \cancel{\text{min}}} \cdot 10\ \cancel{\text{min}} = 50\ \text{gal}$$

Name _____ Date _____

Alejandra drove from Michigan to Colorado to visit her friend. The speed limit on the highway is 70 miles/hour. If Alejandra's combined driving time for the trip was 14 hours, how many miles did Alejandra drive?

1. Ami earns $15 per hour working at the local greenhouse. If she worked 13 hours this month, how much money did she make this month?

$$\frac{15 \text{ dollars}}{1 \text{ hour}} \cdot 13 \text{ hours} = 15 \text{ dollars} \cdot 13 = 195 \text{ dollars}$$

> I know the rate of Ami's pay is 15 dollars for every 1 hour, or $15 \frac{\text{dollars}}{\text{hour}}$. I can multiply the amount of hours to this rate to determine the amount of dollars she makes this month.

At a rate of $15 \frac{\text{dollars}}{\text{hour}}$, Ami will make $195 if she works 13 hours.

2. Trisha is filling her pool. Her pool holds 18,000 gallons of water. The hose she is filling the pool with pumps water at a rate of $300 \frac{\text{gallors}}{\text{hour}}$. If she wants to open her pool in 72 hours, will the pool be full in time?

$$300 \frac{\text{gallons}}{\text{hour}} \cdot 72 \text{ hours} = 300 \text{ gallons} \cdot 72 = 21,600 \text{ gallons}$$

Trisha has plenty of time to fill her pool at this rate. It takes 72 hours to fill 21,600 gallons. She only needs to fill 18,000 gallons.

1. Enguun earns $17 per hour tutoring student-athletes at Brooklyn University.

    a. If Enguun tutored for 12 hours this month, how much money did she earn this month?

    b. If Enguun tutored for 19.5 hours last month, how much money did she earn last month?

2. The Piney Creek Swim Club is preparing for the opening day of the summer season. The pool holds 22,410 gallons of water, and water is being pumped in at 540 gallons per hour. The swim club has its first practice in 42 hours. Will the pool be full in time? Explain your answer.

Analyze tables, graphs, and equations in order to compare rates.

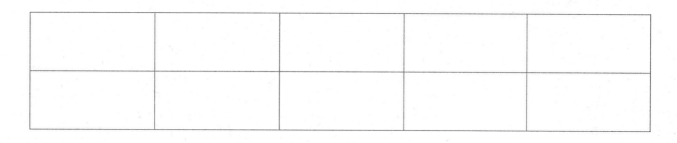

Examples: Creating Tables from Equations

1. The ratio of cups of blue paint to cups of red paint is $1:2$, which means for every cup of blue paint, there are two cups of red paint. In this case, the equation would be red $= 2 \times$ blue, or $r = 2b$, where b represents the amount of blue paint and r represents the amount of red paint. Make a table of values.

2. Ms. Siple is a librarian who really enjoys reading. She can read $\dfrac{3}{4}$ of a book in one day. This relationship can be represented by the equation $b = \dfrac{3}{4}d$, where b represents the number of books and d represents the number of days.

Exercises

1. Bryan and ShaNiece are both training for a bike race and want to compare who rides his or her bike at a faster rate. Both bikers use apps on their phones to record the time and distance of their bike rides. Bryan's app keeps track of his route on a table, and ShaNiece's app presents the information on a graph. The information is shown below.

Bryan:

Number of Hours	0	3	6
Number of Miles	0	75	150

ShaNiece:

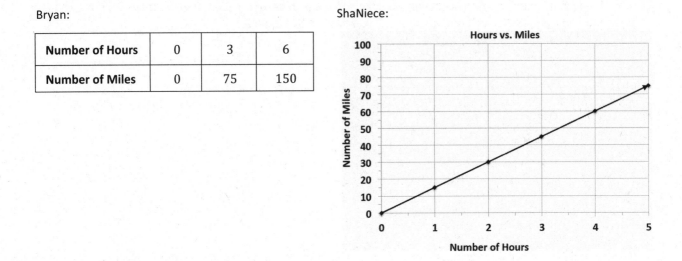

a. At what rate does each biker travel? Explain how you arrived at your answer.

b. ShaNiece wants to win the bike race. Make a new graph to show the speed ShaNiece would have to ride her bike in order to beat Bryan.

Lesson 19:   Comparison Shopping—Unit Price and Related Measurement Conversions

2. Braylen and Tyce both work at a department store and are paid by the hour. The manager told the boys they both earn the same amount of money per hour, but Braylen and Tyce did not agree. They each kept track of how much money they earned in order to determine if the manager was correct. Their data is shown below.

Braylen: $m = 10.50h$ where h represents the number of hours worked and m represents the amount of money Braylen was paid.

Tyce:

Number of Hours	0	3	6
Money in Dollars	0	34.50	69

a. How much did each person earn in one hour?

b. Was the manager correct? Why or why not?

3. Claire and Kate are entering a cup stacking contest. Both girls have the same strategy: stack the cups at a constant rate so that they do not slow down at the end of the race. While practicing, they keep track of their progress, which is shown below.

Claire:

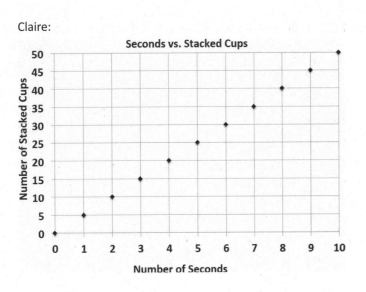

Kate: $c = 4t$, where t represents the amount of time in seconds and c represents the number of stacked cups.

a. At what rate does each girl stack her cups during the practice sessions?

b. Kate notices that she is not stacking her cups fast enough. What would Kate's equation look like if she wanted to stack cups faster than Claire?

Lesson 19:   Comparison Shopping—Unit Price and Related Measurement Conversions

Lesson Summary

When comparing rates and ratios, it is best to find the unit rate.

Comparing unit rates can happen across tables, graphs, and equations.

Name _____   Date _____

Kiara, Giovanni, and Ebony are triplets and always argue over who can answer basic math facts the fastest. After completing a few different math fact activities, Kiara, Giovanni, and Ebony record their data, which is shown below.

Kiara: $m = 5t$, where t represents the time in seconds, and m represents the number of math facts completed.

Ebony:

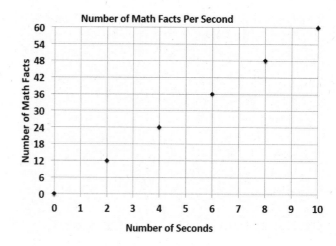

Giovanni:

Seconds	5	10	15
Math Facts	20	40	60

1. What is the math fact completion rate for each student?

2. Who would win the argument? How do you know?

1. Luke is deciding which motorcycle he would like to purchase from the dealership. He has two favorites and will base his final decision on which has the better gas efficiency (the motorcycle that provides more miles for every gallon of gas). The data he received about his first choice, the Trifecta, is represented in the table. The data he received about his second choice, the Zephyr, is represented in the graph. Which motorcycle should Luke purchase?

Trifecta:

Number of Gallons of Gas	3	6	9
Number of Miles	180	360	540

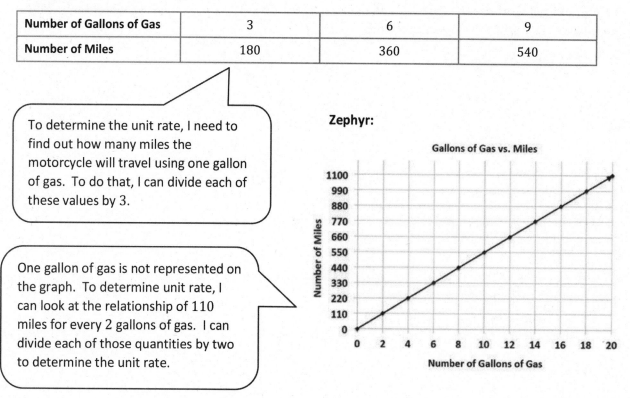

To determine the unit rate, I need to find out how many miles the motorcycle will travel using one gallon of gas. To do that, I can divide each of these values by 3.

One gallon of gas is not represented on the graph. To determine unit rate, I can look at the relationship of 110 miles for every 2 gallons of gas. I can divide each of those quantities by two to determine the unit rate.

Zephyr:

The Trifecta gets $60 \frac{\text{miles}}{\text{gallon}}$ *because* $\frac{180}{3} = 60$. *The Zephyr gets* $55 \frac{\text{miles}}{\text{gallon}}$ *because* $\frac{110}{2} = 55$. *Luke should purchase the Trifecta because it gets more miles for every gallon of gas.*

2. Just as Luke made his final decision, the dealer suggested purchasing the Comet, which gets 928 miles for every tank fill up. The gas tank holds 16 gallons of gas. Is the Comet Luke's best choice, based on miles per gallon?

$$\frac{928}{16}\frac{\text{miles}}{\text{gallon}} = 58\frac{\text{miles}}{\text{gallon}}$$

Luke should still purchase the Trifecta. The Comet only provides 58 miles per gallon, which is less than the 60 miles per gallon the Trifecta provides.

Lesson 19:   Comparison Shopping—Unit Price and Related Measurement Conversions

Victor was having a hard time deciding which new vehicle he should buy. He decided to make the final decision based on the gas efficiency of each car. A car that is more gas efficient gets more miles per gallon of gas. When he asked the manager at each car dealership for the gas mileage data, he received two different representations, which are shown below.

Vehicle 1: Legend

Gallons of Gas	4	8	12
Number of Miles	72	144	216

Vehicle 2: Supreme

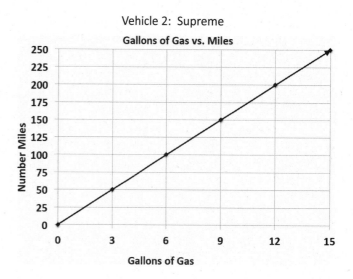

1. If Victor based his decision only on gas efficiency, which car should he buy? Provide support for your answer.

2. After comparing the Legend and the Supreme, Victor saw an advertisement for a third vehicle, the Lunar. The manager said that the Lunar can travel about 289 miles on a tank of gas. If the gas tank can hold 17 gallons of gas, is the Lunar Victor's best option? Why or why not?

An activity will be completed in order to gain confidence in comparing rates on tables, graphs, and equations.

Example 1: Notes from Exit Ticket

Take notes from the discussion in the space provided below.

Notes:

Exploratory Challenge

a. Mallory is on a budget and wants to determine which cereal is a better buy. A 10-ounce box of cereal costs $2.79, and a 13-ounce box of the same cereal costs $3.99.

    i.   Which box of cereal should Mallory buy?

    ii. What is the difference between the two unit prices?

b. Vivian wants to buy a watermelon. Kingston's Market has 10-pound watermelons for $3.90, but the Farmer's Market has 12-pound watermelons for $4.44.

  i. Which market has the best price for watermelon?

  ii. What is the difference between the two unit prices?

c. Mitch needs to purchase soft drinks for a staff party. He is trying to figure out if it is cheaper to buy the 12-pack of soda or the 20-pack of soda. The 12-pack of soda costs $3.99, and the 20-pack of soda costs $5.48.

  i. Which pack should Mitch choose?

  ii. What is the difference in cost between single cans of soda from each of the two packs?

d. Mr. Steiner needs to purchase 60 AA batteries. A nearby store sells a 20-pack of AA batteries for $12.49 and a 12-pack of the same batteries for $7.20.

  i. Would it be less expensive for Mr. Steiner to purchase the batteries in 20-packs or 12-packs?

  ii. What is the difference between the costs of one battery from each pack?

e. The table below shows the amount of calories Mike burns as he runs.

Number of Miles Ran	3	6	9	12
Number of Calories Burned	360	720		1,440

Fill in the missing part of the table.

f.   Emilio wants to buy a new motorcycle. He wants to compare the gas efficiency for each motorcycle before he makes a purchase. The dealerships presented the data below.

Sports Motorcycle:

Number of Gallons of Gas	5	10	15	20
Number of Miles	287.5	575	862.5	1,150

Leisure Motorcycle:

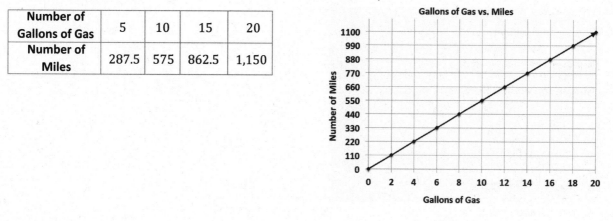

Which motorcycle is more gas efficient and by how much?

g.   Milton Middle School is planning to purchase a new copy machine. The principal has narrowed the choice to two models: SuperFast Deluxe and Quick Copies. He plans to purchase the machine that copies at the fastest rate. Use the information below to determine which copier the principal should choose.

SuperFast Deluxe:

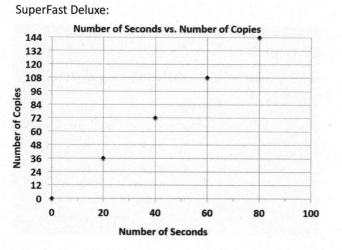

Quick Copies:

$$c = 1.5t$$

(where t represents the amount of time in seconds, and c represents the number of copies)

h.   Elijah and Sean are participating in a walk-a-thon. Each student wants to calculate how much money he would make from his sponsors at different points of the walk-a-thon. Use the information in the tables below to determine which student would earn more money if they both walked the same distance. How much more money would that student earn per mile?

Elijah's Sponsor Plan:

Number of Miles Walked	7	14	21	28
Money Earned in Dollars	35	70	105	140

Sean's Sponsor Plan:

Number of Miles Walked	6	12	18	24
Money Earned in Dollars	33	66	99	132

i.   Gerson is going to buy a new computer to use for his new job and also to download movies. He has to decide between two different computers. How many more kilobytes does the faster computer download in one second?

Choice 1: The rate of download is represented by the following equation: $k = 153t$, where t represents the amount of time in seconds, and k represents the number of kilobytes.

Choice 2: The rate of download is represented by the following equation: $k = 150t$, where t represents the amount of time in seconds, and k represents the number of kilobytes.

      Lesson 20:   Comparison Shopping—Unit Price and Related Measurement Conversions

j. Zyearaye is trying to decide which security system company he will make more money working for. Use the graphs below that show Zyearaye's potential commission rate to determine which company will pay Zyearaye more commission. How much more commission would Zyearaye earn by choosing the company with the better rate?

Superior Security: Top Notch Security:

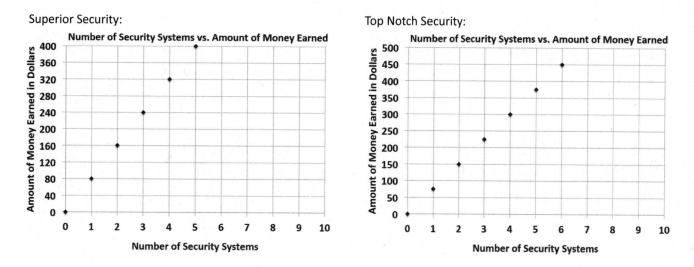

k. Emilia and Miranda are sisters, and their mother just signed them up for a new cell phone plan because they send too many text messages. Using the information below, determine which sister sends the most text messages. How many more text messages does this sister send per week?

Emilia:

Number of Weeks	3	6	9	12
Number of Text Messages	1,200	2,400	3,600	4,800

Miranda: $m = 410w$, where w represents the number of weeks, and m represents the number of text messages.

Lesson Summary

Unit Rate can be located in tables, graphs, and equations.

- Table–the unit rate is the value of the first quantity when the second quantity is **1**.

- Graphs–the unit rate is the value of r at the point $(1, r)$.

- Equation–the unit rate is the constant number in the equation. For example, the unit rate in $r = 3b$ is **3**.

Name _____   Date _____

Value Grocery Mart and Market City are both having a sale on the same popular crackers. McKayla is trying to determine which sale is the better deal. Using the given table and equation, determine which store has the better deal on crackers? Explain your reasoning. (Remember to round your answers to the nearest penny.)

Value Grocery Mart:

Number of Boxes of Crackers	3	6	9	12
Cost (in dollars)	5	10	15	20

Market City:

$c = 1.75b,$ where c represents the cost in dollars, and b represents the number of boxes of crackers.

1. The table below shows how much money Hillary makes working at a yogurt shop. How much money does Hillary make per hour?

Number of Hours Worked	2	4	6	8
Money Earned (in dollars)	25.50	51	76.50	102

> To determine the unit rate, I need to find out how much money Hillary makes in one hour. Since I know how much money she makes in 2 hours, I can divide both of these values by 2.

Hillary earns $\dfrac{25.50}{2} \dfrac{dollars}{hour}$.

$25.50 \div 2 = 12.75$

Hillary earns $12.75 per hour.

2. Makenna is also an employee at the yogurt shop. She earns $2.00 more an hour than Hillary. Complete the table below to show the amount of money Makenna earns.

Number of Hours Worked	3	6	9	12
Money Earned (in dollars)	44.25	88.50	132.75	177

$14.75 \dfrac{dollars}{hour} \cdot 3 \; \text{hours} = 44.25 \; \text{dollars}$

$14.75 \dfrac{dollars}{hour} \cdot 9 \; \text{hours} = 132.75 \; \text{dollars}$

$14.75 \dfrac{dollars}{hour} \cdot 6 \; \text{hours} = 88.50 \; \text{dollars}$

$14.75 \dfrac{dollars}{hour} \cdot 12 \; \text{hours} = 177 \; \text{dollars}$

3. Colbie is also an employee of the yogurt shop. The amount of money she earns is represented by the equation $m = 15h$, where h represents the number of hours worked and m represents the amount of money she earns in dollars. How much more money does Colbie earn an hour than Hillary? Explain your thinking.

The amount of money that Colbie earns for every hour is represented by the constant 15. This tells me that Colbie earns 15 dollars per hour. To determine how much more money an hour she earns than Hillary, I need to subtract Hillary's pay rate from Colbie's pay rate. $15 - 12.75 = 2.25$. Colbie makes 2.25 more dollars per hour than Hillary.

4. Makenna recently received a raise and now makes the same amount of money per hour as Colbie. How much more money per hour does Makenna make now, after her promotion? Explain your thinking.

Makenna now earns the same amount of money per hour as Colbie, which is $15 \frac{\text{dollars}}{\text{hour}}$. She previously earned $14.75 \frac{\text{dollars}}{\text{hour}}$. To determine how much more money Makenna makes now, after her promotion, I need to subtract her previous pay rate from her current pay rate. $15 - 14.75 = 0.25$. Makenna makes 0.25 dollars more an hour, or she makes 25 cents more an hour.

The table below shows the amount of money Gabe earns working at a coffee shop.

Number of Hours Worked	3	6	9	12
Money Earned (in dollars)	40.50	81.00	121.50	162.00

1. How much does Gabe earn per hour?

2. Jordan is another employee at the same coffee shop. He has worked there longer than Gabe and earns $3 more per hour than Gabe. Complete the table below to show how much Jordan earns.

Number of Hours Worked	4	8	12	16
Money Earned (in dollars)				

3. Serena is the manager of the coffee shop. The amount of money she earns is represented by the equation $m = 21h$, where h is the number of hours Serena works, and m is the amount of money she earns. How much more money does Serena make an hour than Gabe? Explain your thinking.

4. Last month, Jordan received a promotion and became a manager. He now earns the same amount as Serena. How much more money does he earn per hour now that he is a manager than he did before his promotion? Explain your thinking.

Conversion tables contain ratios that can be used to convert units of length, weight, or capacity. You must multiply the given number by the ratio that compares the two units.

Opening Exercise

Identify the ratios that are associated with conversions between feet, inches, and yards.

12 inches = _____ foot; the ratio of inches to feet is _____.

1 foot = _____ inches; the ratio of feet to inches is _____.

3 feet = _____ yard; the ratio of feet to yards is _____.

1 yard = _____ feet; the ratio of yards to feet is _____.

Example 1

Work with your partner to find out how many feet are in 48 inches. Make a ratio table that compares feet and inches. Use the conversion rate of 12 inches per foot or $\frac{1}{12}$ foot per inch.

Example 2

How many grams are in 6 kilograms? Again, make a record of your work before using the calculator. The rate would be 1,000 grams per kilogram. The unit rate would be 1,000.

Exercise 1

How many cups are in 5 quarts? As always, make a record of your work before using the calculator. The rate would be 4 cups per quart. The unit rate would be 4.

Exercise 2

How many quarts are in 10 cups?

Lesson 21:   Getting the Job Done—Speed, Work, and Measurement Units

Lesson Summary

Conversion tables contain ratios that can be used to convert units of length, weight, or capacity. You must multiply the given number by the ratio that compares the two units.

U.S. Customary Length	Conversion
Inch (in.)	$1\,\text{in.} = \dfrac{1}{12}\,\text{ft.}$
Foot (ft.)	1 ft. = 12 in.
Yard (yd.)	1 yd. = 3 ft. 1 yd. = 36 in.
Mile (mi.)	1 mi. = 1,760 yd. 1 mi. = 5,280 ft.

Metric Length	Conversion
Centimeter (cm)	1 cm = 10 mm
Meter (m)	1 m = 100 cm 1 m = 1,000 mm
Kilometer (km)	1 km = 1,000 m

U.S. Customary Weight	Conversion
Pound (lb.)	1 lb. = 16 oz.
Ton (T.)	1 T. = 2,000 lb.

Metric Capacity	Conversion
Liter (L)	1 L = 1,000 ml
kiloliter (kL)	1 kL = 1,000 L

U.S. Customary Capacity	Conversion
Cup (c.)	1 c. = 8 fluid ounces
pint (pt.)	1 pt. = 2 c.
Quart (qt.)	1 qt. = 4 c. 1 qt. = 2 pt. 1 qt. = 32 fluid ounces
Gallon (gal.)	1 gal. = 4 qt. 1 gal. = 8 pt. 1 gal. = 16 c. 1 gal. = 128 fluid ounces

Metric Mass	Conversion
Gram (g)	1 g = 1,000 mg
kilogram (kg)	1 kg = 1,000 g

Name _____   Date _____

Jill and Erika made 4 gallons of lemonade for their lemonade stand. How many quarts did they make? If they charge $2.00 per quart, how much money will they make if they sell it all?

1. 4 km = _____ m

   4,000

   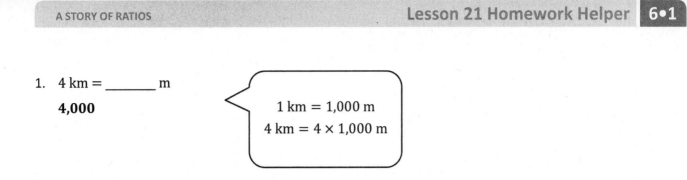

   1 km = 1,000 m

   4 km = 4 × 1,000 m

2. Matt buys 2 pounds of popcorn. He will give each friend a one-ounce bag of popcorn. How many bags can Matt make?

   32 *bags*

   1 lb. = 16 oz.

   2 lb. = 2 × 16 oz.

1. 7 ft. = _____ in.

2. 100 yd. = _____ ft.

3. 25 m = _____ cm

4. 5 km = _____ m

5. 96 oz. = _____ lb.

6. 2 mi. = _____ ft.

7. 2 mi. = _____ yd.

8. 32 fl. oz. = _____ c.

9. 1,500 mL = _____ L

10. 6 g = _____ mg

11. Beau buys a 3-pound bag of trail mix for a hike. He wants to make one-ounce bags for his friends with whom he is hiking. How many one-ounce bags can he make? _____

12. The maximum weight for a truck on the New York State Thruway is 40 tons. How many pounds is this? _____

13. Claudia's skis are 150 centimeters long. How many meters is this? _____

14. Claudia's skis are 150 centimeters long. How many millimeters is this? _____

15. Write your own problem, and solve it. Be ready to share the question tomorrow.

If an object is moving at a constant rate of speed for a certain amount of time, it is possible to find how far the object went by multiplying the rate and the time. In mathematical language, we say, distance = rate · time.

Example 1

Walker: Substitute the walker's distance and time into the equation and solve for the rate of speed.

distance = rate · time

$d = r \cdot t$

Hint: Consider the units that you want to end up with. If you want to end up with the rate (feet/second), then divide the distance (feet) by time (seconds).

Runner: Substitute the runner's time and distance into the equation to find the rate of speed.

distance = rate · time

$d = r \cdot t$

Lesson 22:   Getting the Job Done—Speed, Work, and Measurement Units

Example 2

Part 1: Chris Johnson ran the 40-yard dash in 4.24 seconds. What is the rate of speed? Round any answer to the nearest hundredth.

distance $=$ rate $\cdot$ time

$d = r \cdot t$

Part 2: In Lesson 21, we converted units of measure using unit rates. If the runner were able to run at a constant rate, how many yards would he run in an hour? This problem can be solved by breaking it down into two steps. Work with a partner, and make a record of your calculations.

a.   How many yards would he run in one minute?

b.   How many yards would he run in one hour?

We completed that problem in two separate steps, but it is possible to complete this same problem in one step. We can multiply the yards per second by the seconds per minute, then by the minutes per hour.

$$\underline{\hspace{2cm}}\,\frac{\text{yards}}{\text{second}} \cdot 60\,\frac{\text{seconds}}{\text{minute}} \cdot 60\,\frac{\text{minutes}}{\text{hour}} = \underline{\hspace{3cm}}\ \text{yards in one hour}$$

Cross out any units that are in both the numerator and denominator in the expression because these cancel each other out.

Part 3: How many miles did the runner travel in that hour? Round your response to the nearest tenth.

Cross out any units that are in both the numerator and denominator in the expression because they cancel out.

Exercises: Road Trip

Exercise 1

I drove my car on cruise control at 65 miles per hour for 3 hours without stopping. How far did I go?

$d = r \cdot t$

$d = \underline{\hspace{2cm}} \dfrac{\text{miles}}{\text{hour}} \cdot \underline{\hspace{1.5cm}} \text{ hours}$

Cross out any units that are in both the numerator and denominator in the expression because they cancel out.

$d = \underline{\hspace{2cm}} \text{ miles}$

Exercise 2

On the road trip, the speed limit changed to 50 miles per hour in a construction zone. Traffic moved along at a constant rate (50 mph), and it took me 15 minutes (0.25 hours) to get through the zone. What was the distance of the construction zone? (Round your response to the nearest hundredth of a mile.)

$d = r \cdot t$

$d = \underline{\hspace{2cm}} \dfrac{\text{miles}}{\text{hour}} \cdot \underline{\hspace{1.5cm}} \text{ hours}$

Lesson Summary

Distance, rate, and time are related by the formula $d = r \cdot t$.

Knowing any two of the values allows the calculation of the third.

Name _____ Date _____

Franny took a road trip to her grandmother's house. She drove at a constant speed of 60 miles per hour for 2 hours. She took a break and then finished the rest of her trip driving at a constant speed of 50 miles per hour for 2 hours. What was the total distance of Franny's trip?

1. A biplane travels at a constant speed of $500 \ \frac{\text{kilometers}}{\text{hour}}$. It travels at this rate for 2 hours. How far did the biplane travel in this time?

$$500 \ \frac{\text{kilometers}}{\text{hour}} \cdot 2 \ \text{hours} = 500 \ \text{kilometers} \cdot 2 = 1{,}000 \ \text{kilometers}$$

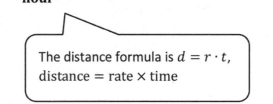

The distance formula is $d = r \cdot t$, distance = rate × time

2. Tina ran a 50 yard race in 5.5 seconds. What is her rate of speed?

$$\frac{50}{5.5} \ \frac{\text{yards}}{\text{second}} = 9.09 \ \frac{\text{yards}}{\text{second}}$$

$r = \frac{d}{t}$

1. If Adam's plane traveled at a constant speed of 375 miles per hour for six hours, how far did the plane travel?

2. A Salt March Harvest Mouse ran a 360 centimeter straight course race in 9 seconds. How fast did it run?

3. Another Salt Marsh Harvest Mouse took 7 seconds to run a 350 centimeter race. How fast did it run?

4. A slow boat to China travels at a constant speed of 17.25 miles per hour for 200 hours. How far was the voyage?

5. The Sopwith Camel was a British, First World War, single-seat, biplane fighter introduced on the Western Front in 1917. Traveling at its top speed of 110 mph in one direction for $2\frac{1}{2}$ hours, how far did the plane travel?

6. A world-class marathon runner can finish 26.2 miles in 2 hours. What is the rate of speed for the runner?

7. Banana slugs can move at 17 cm per minute. If a banana slug travels for 5 hours, how far will it travel?

- If work is being done at a constant rate by one person, and at a different constant rate by another person, both rates can be converted to their unit rates and then compared directly.

- "Work" can include jobs done in a certain time period, rates of running or swimming, etc.

Example 1: Fresh-Cut Grass

Suppose that on a Saturday morning you can cut 3 lawns in 5 hours, and your friend can cut 5 lawns in 8 hours. Who is cutting lawns at a faster rate?

$$\frac{3 \text{ lawns}}{5 \text{ hours}} = \frac{\underline{\hspace{1cm}} \text{ lawns}}{1 \text{ hour}} \qquad\qquad \frac{5 \text{ lawns}}{8 \text{ hours}} = \frac{\underline{\hspace{1cm}} \text{ lawns}}{1 \text{ hour}}$$

Example 2: Restaurant Advertising

$$\frac{\underline{\hspace{0.7cm}} \text{ menus}}{\underline{\hspace{0.7cm}} \text{ hours}} = \frac{\underline{\hspace{0.7cm}} \text{ menus}}{1 \text{ hour}} \qquad\qquad \frac{\underline{\hspace{0.7cm}} \text{ menus}}{\underline{\hspace{0.7cm}} \text{ hours}} = \frac{\underline{\hspace{0.7cm}} \text{ menus}}{1 \text{ hour}}$$

Example 3: Survival of the Fittest

$$\frac{\underline{\qquad}}{\underline{\qquad}} \frac{\text{feet}}{\text{seconds}} = \frac{\underline{\qquad}}{1} \frac{\text{feet}}{\text{second}} \qquad\qquad \frac{\underline{\qquad}}{\underline{\qquad}} \frac{\text{feet}}{\text{seconds}} = \frac{\underline{\qquad}}{1} \frac{\text{feet}}{\text{second}}$$

Example 4: Flying Fingers

$$\underline{\qquad\qquad} = \underline{\qquad\qquad} \qquad\qquad\qquad \underline{\qquad\qquad} = \underline{\qquad\qquad}$$

Lesson Summary

- Rate problems, including constant rate problems, always count or measure something happening per unit of time. The time is always in the denominator.

- Sometimes the units of time in the denominators of the rates being compared are not the same. One must be converted to the other before calculating the unit rate of each.

Name _____ Date _____

A sixth-grade math teacher can grade 25 homework assignments in 20 minutes.

Is he working at a faster rate or slower rate than grading 36 homework assignments in 30 minutes?

Who runs at a faster rate: someone who runs 40 yards in 5.8 seconds or someone who runs 100 yards in 10 seconds?

$$\frac{40}{5.8}\ \frac{\text{yards}}{\text{second}} \approx 6.9\ \frac{\text{yards}}{\text{second}}$$

$$\frac{100}{10}\ \frac{\text{yards}}{\text{second}} = 10\ \frac{\text{yards}}{\text{second}} \rightarrow \text{faster}$$

Find the unit rate by dividing. Compare the unit rates to determine the faster runner.

1. Who walks at a faster rate: someone who walks 60 feet in 10 seconds or someone who walks 42 feet in 6 seconds?

2. Who walks at a faster rate: someone who walks 60 feet in 10 seconds or someone who takes 5 seconds to walk 25 feet? Review the lesson summary before answering.

3. Which parachute has a slower descent: a red parachute that falls 10 feet in 4 seconds or a blue parachute that falls 12 feet in 6 seconds?

4. During the winter of 2012–2013, Buffalo, New York received 22 inches of snow in 12 hours. Oswego, New York received 31 inches of snow over a 15 hour period. Which city had a heavier snowfall rate? Round your answers to the nearest hundredth.

5. A striped marlin can swim at a rate of 70 miles per hour. Is this a faster or slower rate than a sailfish, which takes 30 minutes to swim 40 miles?

6. One math student, John, can solve 6 math problems in 20 minutes while another student, Juaquine, can solve the same 6 math problems at a rate of 1 problem per 4 minutes. Who works faster?

Exercise 1

Robb's Fruit Farm consists of 100 acres on which three different types of apples grow. On 25 acres, the farm grows Empire apples. McIntosh apples grow on 30% of the farm. The remainder of the farm grows Fuji apples. Shade in the grid below to represent the portion of the farm each type of apple occupies. Use a different color for each type of apple. Create a key to identify which color represents each type of apple.

Color Key

Empire _____

McIntosh _____

Fuji _____

Part-to-Whole Ratio

_____

_____

_____

Exercise 2

The shaded portion of the grid below represents the portion of a granola bar remaining.

What percent does each block of granola bar represent?

What percent of the granola bar remains?

What other ways can we represent this percent?

0.01	0.01	0.01	0.01	0.01	0.01	0.01	0.01	0.01	0.01
0.01	0.01	0.01	0.01	0.01	0.01	0.01	0.01	0.01	0.01
0.01	0.01	0.01	0.01	0.01	0.01	0.01	0.01	0.01	0.01
0.01	0.01	0.01	0.01	0.01	0.01	0.01	0.01	0.01	0.01
0.01	0.01	0.01	0.01	0.01	0.01	0.01	0.01	0.01	0.01
0.01	0.01	0.01	0.01	0.01	0.01	0.01	0.01	0.01	0.01
0.01	0.01	0.01	0.01	0.01	0.01	0.01	0.01	0.01	0.01
0.01	0.01	0.01	0.01	0.01	0.01	0.01	0.01	0.01	0.01
0.01	0.01	0.01	0.01	0.01	0.01	0.01	0.01	0.01	0.01
0.01	0.01	0.01	0.01	0.01	0.01	0.01	0.01	0.01	0.01

Exercise 3

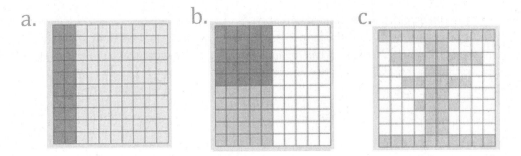

a.   For each figure shown, represent the gray shaded region as a percent of the whole figure.   Write your answer as a decimal, fraction, and percent.

Picture (a)	Picture (b)	Picture (c)

b.   What ratio is being modeled in each picture?

c.   How are the ratios and percentages related?

Lesson 24:   Percent and Rates per 100

Exercise 4

Each relationship below compares the shaded portion (the part) to the entire figure (the whole). Complete the table.

Percentage	Decimal	Fraction	Ratio	Model
6%			6 : 100	
60%				
600%				
32%				

	0.55			
		$\dfrac{9}{10}$		

Exercise 5

Mr. Brown shares with the class that 70% of the students got an A on the English vocabulary quiz. If Mr. Brown has 100 students, create a model to show how many of the students received an A on the quiz.

What fraction of the students received an A on the quiz?

How could we represent this amount using a decimal?

How are the decimal, the fraction, and the percent all related?

Exercise 6

Marty owns a lawn mowing service. His company, which consists of three employees, has 100 lawns to mow this week. Use the 10×10 grid to model how the work could have been distributed between the three employees.

Worker	Percentage	Fraction	Decimal
Employee 1			
Employee 2			
Employee 3			

Color over the name with the same color you used in the diagram.

Lesson Summary

One percent is the number $\frac{1}{100}$ and is written as 1%.

Percentages can be used as rates. For example, 30% of a quantity means $\frac{30}{100}$ times the quantity.

We can create models of percents. One example would be to shade a 10×10 grid. Each square in a 10×10 grid represents 1% or 0.01.

Name _____ Date _____

One hundred offices need to be painted. The workers choose between yellow, blue, or beige paint. They decide that 45% of the offices will be painted yellow; 28% will be painted blue, and the remaining offices will be painted beige. Create a model that shows the percent of offices that will be painted each color. Write the amounts as decimals and fractions.

Color	%	Fraction	Decimal
Yellow			
Blue			
Beige			

1. Holly owns a home cleaning service. Her company, Holly N'Helpers, which consists of three employees, has 100 homes to clean this month. Use the 10 × 10 grid to model how the work could have been distributed between the three employees. Using your model, complete the table.

   Answers can vary as students choose how they want to separate the workload. This is a sample response.

B	B	G	G	G	G	G	P	P	P
B	B	G	G	G	G	G	P	P	P
B	B	G	G	G	G	G	P	P	P
B	B	G	G	G	G	G	P	P	P
B	B	G	G	G	G	G	P	P	P
B	B	B	G	G	G	G	P	P	P
B	B	B	G	G	G	G	P	P	P
B	B	B	G	G	G	G	P	P	P
B	B	B	G	G	G	G	P	P	P
B	B	B	G	G	G	G	P	P	P

Worker	Percentage	Fraction	Decimal
Employee B	25%	$\dfrac{25}{100}$	0.25
Employee G	45%	$\dfrac{45}{100}$	0.45
Employee P	30%	$\dfrac{30}{100}$	0.30

I will assign Employee B to clean 25 houses, Employee G to clean 45 houses, and Employee P to clean 30 houses.

I know percents are out of a total of 100 and are another way to show a part–to–whole ratio. Since there are 100 houses to clean, the total is 100 in this example. Since Employee B is assigned to 25 homes, the ratio is $25 : 100$, the fraction is $\dfrac{25}{100}$, the decimal is 0.25, and the percentage is 25%. Using this reasoning, I was able to complete the table for Employees G and P.

2. When hosting Math Carnival at the middle school, 80 percent of the budget is spent on prizes for the winners of each game. Shade the grid below to represent the portion of the budget that is spent on prizes.

0.01	0.01	0.01	0.01	0.01	0.01	0.01	0.01	0.01	0.01
0.01	0.01	0.01	0.01	0.01	0.01	0.01	0.01	0.01	0.01
0.01	0.01	0.01	0.01	0.01	0.01	0.01	0.01	0.01	0.01
0.01	0.01	0.01	0.01	0.01	0.01	0.01	0.01	0.01	0.01
0.01	0.01	0.01	0.01	0.01	0.01	0.01	0.01	0.01	0.01
0.01	0.01	0.01	0.01	0.01	0.01	0.01	0.01	0.01	0.01
0.01	0.01	0.01	0.01	0.01	0.01	0.01	0.01	0.01	0.01
0.01	0.01	0.01	0.01	0.01	0.01	0.01	0.01	0.01	0.01
0.01	0.01	0.01	0.01	0.01	0.01	0.01	0.01	0.01	0.01
0.01	0.01	0.01	0.01	0.01	0.01	0.01	0.01	0.01	0.01

> This is a 10×10 grid, so I know there are 100 total blocks. I know each block represents $\frac{1}{100}$, 0.01, or 1%. 80% means 80 out of 100, so I will shade 80 blocks.

a. What does each block represent?

Each block represents $\frac{1}{100}$ of the total budget.

> Because there are 100 total blocks, which represents the entire budget, each block represents $\frac{1}{100}$ of the total budget.

b. What percent of the budget was not spent on prizes?

20%

> I know from the problem that 80% of the total budget was spent on prizes, so this leaves 20% of the budget for other expenses. $100\% - 80\% = 20\%$

1. Marissa just bought 100 acres of land. She wants to grow apple, peach, and cherry trees on her land. Color the model to show how the acres could be distributed for each type of tree. Using your model, complete the table.

Tree	Percentage	Fraction	Decimal
Apple			
Peach			
Cherry			

2. After renovations on Kim's bedroom, only 30 percent of one wall is left without any décor. Shade the grid below to represent the space that is left to decorate.

   a. What does each block represent?

   b. What percent of this wall has been decorated?

0.01	0.01	0.01	0.01	0.01	0.01	0.01	0.01	0.01	0.01
0.01	0.01	0.01	0.01	0.01	0.01	0.01	0.01	0.01	0.01
0.01	0.01	0.01	0.01	0.01	0.01	0.01	0.01	0.01	0.01
0.01	0.01	0.01	0.01	0.01	0.01	0.01	0.01	0.01	0.01
0.01	0.01	0.01	0.01	0.01	0.01	0.01	0.01	0.01	0.01
0.01	0.01	0.01	0.01	0.01	0.01	0.01	0.01	0.01	0.01
0.01	0.01	0.01	0.01	0.01	0.01	0.01	0.01	0.01	0.01
0.01	0.01	0.01	0.01	0.01	0.01	0.01	0.01	0.01	0.01
0.01	0.01	0.01	0.01	0.01	0.01	0.01	0.01	0.01	0.01
0.01	0.01	0.01	0.01	0.01	0.01	0.01	0.01	0.01	0.01

Example 1

Sam says 50% of the vehicles are cars. Give three different reasons or models that prove or disprove Sam's statement. Models can include tape diagrams, 10×10 grids, double number lines, etc.

How is the fraction of cars related to the percent?

Use a model to prove that the fraction and percent are equivalent.

What other fractions or decimals also represent 60%?

Example 2

A survey was taken that asked participants whether or not they were happy with their job. An overall score was given. 300 of the participants were unhappy while 700 of the participants were happy with their job. Give a part-to-whole fraction for comparing happy participants to the whole. Then write a part-to-whole fraction of the unhappy participants to the whole. What percent were happy with their job, and what percent were unhappy with their job?

Happy   _____   _____   Unhappy   _____   _____

             Fraction   Percent   Fraction   Percent

Create a model to justify your answer.

Exercise 1

Renita claims that a score of 80% means that she answered $\frac{4}{5}$ of the problems correctly. She drew the following picture to support her claim:

Is Renita correct? _____ Why or why not?

How could you change Renita's picture to make it easier for Renita to see why she is correct or incorrect?

Exercise 2

Use the diagram to answer the following questions.

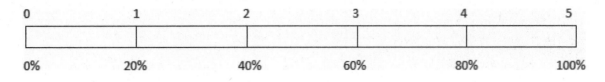

80% is what fraction of the whole quantity?

$\frac{1}{5}$ is what percent of the whole quantity?

50% is what fraction of the whole quantity?

1 is what percent of the whole quantity?

Exercise 3

Maria completed $\frac{3}{4}$ of her workday. Create a model that represents what percent of the workday Maria has worked.

What percent of her workday does she have left?

How does your model prove that your answer is correct?

Exercise 4

Matthew completed $\frac{5}{8}$ of his workday. What decimal would also describe the portion of the workday he has finished?

How can you use the decimal to get the percent of the workday Matthew has completed?

Exercise 5

Complete the conversions from fraction to decimal to percent.

Fraction	Decimal	Percent
$\frac{1}{8}$		
	0.35	
		84.5%
	0.325	
$\frac{2}{25}$		

Exercise 6

Choose one of the rows from the conversion table in Exercise 5, and use models to prove your answers. (Models could include a 10×10 grid, a tape diagram, a double number line, etc.)

Lesson Summary

Fractions, decimals, and percentages are all related.

To change a fraction to a percentage, you can scale up or scale down so that 100 is in the denominator.

Example:

$$\frac{9}{20} = \frac{9 \times 5}{20 \times 5} = \frac{45}{100} = 45\%$$

There may be times when it is more beneficial to convert a fraction to a percent by first writing the fraction in decimal form.

Example:

$$\frac{5}{8} = 0.625 = 62.5 \text{ hundredths} = 62.5\%$$

Models, like tape diagrams and number lines, can also be used to model the relationships.

0%	25%	50%	75%	100%
0	20	40	60	80

The diagram shows that $\frac{20}{80} = 25\%$.

10 × 10 Grid Reproducible

Name _____ Date _____

1. Convert 0.3 to a fraction and a percent.

2. Convert 9% to a fraction and a decimal.

3. Convert $\frac{3}{8}$ to a decimal and a percent.

1. Use the 10 × 10 grid to express the fraction $\frac{7}{25}$ as a percent.

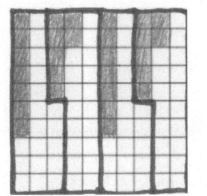

Students should shade 28 of the squares in the grid.

There are 100 squares in the grid. Percent is a part–to–whole comparison where the whole is 100. The fraction is $\frac{7}{25}$, so I will divide the whole (100 squares) into 4 parts since $100 \div 25 = 4$. I can shade 7 squares in each part as seen in the first grid because the fraction $\frac{7}{25}$ tells me there are 7 shaded squares for every group of 25. Since there are 4 groups of 25 in 100, I can also multiply 7 by 4, which will give me the total number of shaded squares, 28.

2. Use a **model** to relate the fraction $\frac{7}{25}$ to a **percent**.

0	7	14	21	28

0	25	50	75	100

3. How are the diagrams related?

    Both show that $\frac{7}{25}$ is equal to $\frac{28}{100}$.

Both grids show that $\frac{7}{25}$ is equal to $\frac{28}{100}$. The model also shows the fractions are the same.

4. What **decimal** is also related to the **fraction**?

    0.28

5. Which diagram is the most helpful for converting the fraction to a decimal? Explain why.

    Answers will vary according to student preferences. Possible student response: It is most helpful to use the tape diagram for converting the fraction to a decimal. To convert $\frac{7}{25}$ to a decimal, I can clearly see how there are 4 groups of 25 in 100. So, to find 4 groups of 7, I can multiply 4×7, which is 28. I know $\frac{7}{25}$ is equal to $\frac{28}{100}$, which is 0.28.

Lesson 25:   A Fraction as a Percent

1. Use the 10×10 grid to express the fraction $\frac{11}{20}$ as a percent.

2. Use a tape diagram to relate the fraction $\frac{11}{20}$ to a percent.

3. How are the diagrams related?

4. What decimal is also related to the fraction?

5. Which diagram is the most helpful for converting the fraction to a decimal? _____ Explain why.

Example 1

Five of the 25 girls on Alden Middle School's soccer team are seventh-grade students. Find the percentage of seventh graders on the team. Show two different ways of solving for the answer. One of the methods must include a diagram or picture model.

Example 2

Of the 25 girls on the Alden Middle School soccer team, 40% also play on a travel team. How many of the girls on the middle school team also play on a travel team?

Example 3

The Alden Middle School girls' soccer team won 80% of its games this season. If the team won 12 games, how many games did it play? Solve the problem using at least two different methods.

Exercises

1. There are 60 animal exhibits at the local zoo. What percent of the zoo's exhibits does each animal class represent?

Exhibits by Animal Class	Number of Exhibits	Percent of the Total Number of Exhibits
Mammals	30	
Reptiles & Amphibians	15	
Fish & Insects	12	
Birds	3	

2. A sweater is regularly $32. It is 25% off the original price this week.

   a. Would the amount the shopper saved be considered the part, whole, or percent?

   b. How much would a shopper save by buying the sweater this week? Show two methods for finding your answer.

3. A pair of jeans was 30% off the original price. The sale resulted in a $24 discount.

    a.   Is the original price of the jeans considered the whole, part, or percent?

    b.   What was the original cost of the jeans before the sale? Show two methods for finding your answer.

4. Purchasing a TV that is 20% off will save $180.

    a.   Name the different parts with the words: PART, WHOLE, PERCENT.

    _____   _____   _____
         20% off $180   Original Price

    b.   What was the original price of the TV? Show two methods for finding your answer.

Name _____   Date _____

1. Find 40% of 60 using two different strategies, one of which must include a pictorial model or diagram.

2. 15% of an amount is 30. Calculate the whole amount using two different strategies, one of which must include a pictorial model.

1. What is 15% of 80? Create a model to prove your answer.

I know the whole, 100%, is 80.

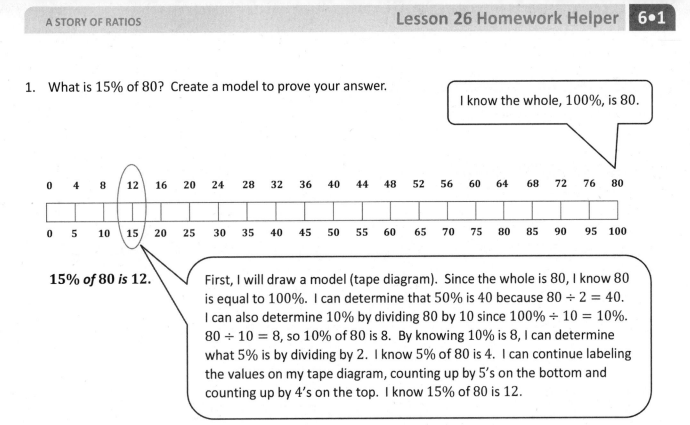

15% *of* 80 *is* 12.

First, I will draw a model (tape diagram). Since the whole is 80, I know 80 is equal to 100%. I can determine that 50% is 40 because $80 \div 2 = 40$. I can also determine 10% by dividing 80 by 10 since $100\% \div 10 = 10\%$. $80 \div 10 = 8$, so 10% of 80 is 8. By knowing 10% is 8, I can determine what 5% is by dividing by 2. I know 5% of 80 is 4. I can continue labeling the values on my tape diagram, counting up by 5's on the bottom and counting up by 4's on the top. I know 15% of 80 is 12.

2. If 30% of a number is 84, what was the original number?

We know from the problem that 30% of a number is 84, so I will label this on the tape diagram.

This is the value I will be determining.

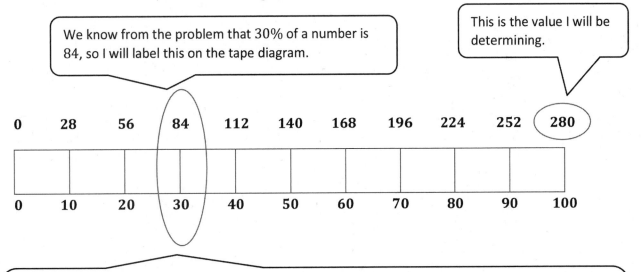

Because the whole represents 100%, I can divide the tape diagram into 10 parts so each part represents 10%. I know 30% is 84. Since I know 30% is 84, I can find 10% by dividing 84 by 3 since $30\% \div 3 = 10\%$. $84 \div 3 = 28$, so 10% is 28. Now that I know the value of 10%, I can determine the value of 100% by multiplying $28 \times 10 = 280$. So, the value of the whole (the original number) is 280.

3. In a 10×10 grid that represents 500, one square represents _____**5**_____.

Use the grid below to represent 17% of 500.

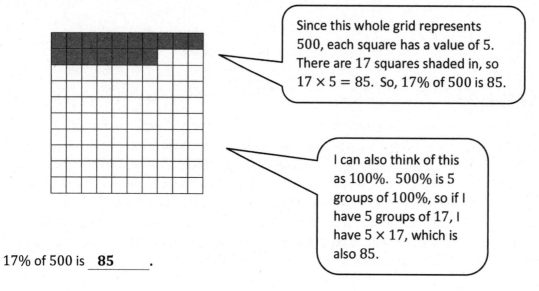

Since this whole grid represents 500, each square has a value of 5. There are 17 squares shaded in, so $17 \times 5 = 85$. So, 17% of 500 is 85.

I can also think of this as 100%. 500% is 5 groups of 100%, so if I have 5 groups of 17, I have 5×17, which is also 85.

17% of 500 is ___**85**_____.

Lesson 26:   Percent of a Quantity

1. What is 15% of 60? Create a model to prove your answer.

2. If 40% of a number is 56, what was the original number?

3. In a 10 × 10 grid that represents 800, one square represents _____.
   Use the grids below to represent 17% and 83% of 800.

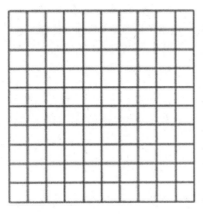

17%

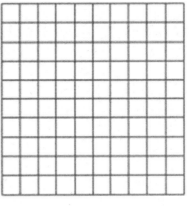

83%

17% of 800 is _____.

83% of 800 is _____.

Example 1

Solve the following three problems.

Write the words PERCENT, WHOLE, or PART under each problem to show which piece you were solving for.

60% of 300 = _____   60% of _____ = 300   60 out of 300 = _____%

How did your solving method differ with each problem?

Exercise

Use models, such as 10×10 grids, ratio tables, tape diagrams, or double number line diagrams, to solve the following situation.

Priya is doing her back-to-school shopping. Calculate all of the missing values in the table below, rounding to the nearest penny, and calculate the total amount Priya will spend on her outfit after she receives the indicated discounts.

	Shirt (25% discount)	Pants (30% discount)	Shoes (15% discount)	Necklace (10% discount)	Sweater (20% discount)
Original Price	$44			$20	
Amount of Discount		$15	$9		$7

What is the total cost of Priya's outfit?

Lesson Summary

Percent problems include the part, whole, and percent. When one of these values is missing, we can use tables, diagrams, and models to solve for the missing number.

Name _____ Date _____

Jane paid $40 for an item after she received a 20% discount. Jane's friend says this means that the original price of the item was $48.

    a. How do you think Jane's friend arrived at this amount?

    b. Is her friend correct? Why or why not?

1. The Soccer Club of Mathematica County is hosting its annual buffet. 40 players are attending this event. 28 players have either received their food or are currently in line. The rest are patiently waiting to be called to the buffet line. What percent of the players are waiting?

$$\frac{12}{40} = \frac{30}{100} = 30\%$$

30% of the players are waiting.

Since 28 players have either received their food or are in line, I will subtract this from 40 to find out how many players are still waiting. $40 - 28 = 12$. I know 12 is a part of 40 (the whole).

$$\frac{12}{40} = \frac{3}{10} = \frac{30}{100} = 30\%$$

2. Dry Clean USA has finished cleaning 25% of their 724 orders. How many orders do they still need to finish cleaning?

They cleaned 181 orders, so they still have 543 orders left to clean.

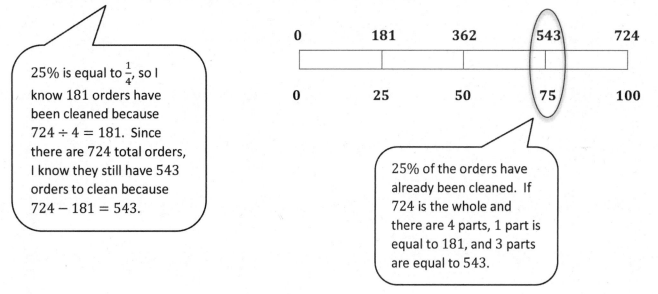

25% is equal to $\frac{1}{4}$, so I know 181 orders have been cleaned because $724 \div 4 = 181$. Since there are 724 total orders, I know they still have 543 orders to clean because $724 - 181 = 543$.

25% of the orders have already been cleaned. If 724 is the whole and there are 4 parts, 1 part is equal to 181, and 3 parts are equal to 543.

1. Mr. Yoshi has 75 papers. He graded 60 papers, and he had a student teacher grade the rest. What percent of the papers did each person grade?

2. Mrs. Bennett has graded 20% of her 150 students' papers. How many papers does she still need to finish grading?

Example

If an item is discounted 20%, the sale price is what percent of the original price?

If the original price of the item is $400, what is the dollar amount of the discount?

How much is the sale price?

Exercise

The following items were bought on sale. Complete the missing information in the table.

Item	Original Price	Sale Price	Amount of Discount	Percent Saved	Percent Paid
Television		$800		20%	
Sneakers	$80			25%	
Video Games		$54			90%
MP3 Player		$51.60		40%	
Book			$2.80		80%
Snack Bar		$1.70	$0.30		

Lesson Summary

Percent problems include the part, whole, and percent. When one of these values is missing, we can use tables, diagrams, and models to solve for the missing number.

Name _____   Date _____

1. Write one problem using a dollar amount of $420 and a percent of 40%. Provide the solution to your problem.

2. The sale price of an item is $160 after a 20% discount. What was the original price of the item?

The school fundraiser was a huge success. Ms. Baker's class is in charge of delivering the orders to the students by the end of the day. They delivered 46 orders so far. If this number represents 20% of the total number of orders, how many total orders will Ms. Baker's class have to deliver before the end of the day?

Ms. Baker's class will have to deliver 230 *total orders.*

$$20\% = \frac{20}{100} = \frac{2}{10} = \frac{46}{230}$$

I know 20% is equal to $\frac{20}{100}$, which can be renamed as $\frac{2}{10}$. I know 46 is a part, and I need to find the whole. I will determine a fraction equivalent to $\frac{2}{10}$, or 20%, by multiplying the numerator and denominator by 23 because $46 \div 2 = 23$. $2 \times 23 = 46$ and $10 \times 23 = 230$.

1. The Sparkling House Cleaning Company has cleaned 28 houses this week. If this number represents 40% of the total number of houses the company is contracted to clean, how many total houses will the company clean by the end of the week?

2. Joshua delivered 30 hives to the local fruit farm. If the farmer has paid to use 5% of the total number of Joshua's hives, how many hives does Joshua have in all?

Exploratory Challenge 1

Claim: To find 10% of a number, all you need to do is move the decimal to the left once.

Use at least one model to solve each problem (e.g., tape diagram, table, double number line diagram, 10×10 grid).

a. Make a prediction. Do you think the claim is true or false? _____ Explain why.

b. Determine 10% of 300 . _____

c. Find 10% of 80. _____

d. Determine 10% of 64. _____

e. Find 10% of 5. _____

f. 10% of _____ is 48.

g. 10% of _____ is 6.

h.   Gary read 34 pages of a 340-page book. What percent did he read?

i.   Micah read 16 pages of his book. If this is 10% of the book, how many pages are in the book?

j.   Using the solutions to the problems above, what conclusions can you make about the claim?

Exploratory Challenge 2

Claim: If an item is already on sale, and then there is another discount taken off the new price, this is the same as taking the sum of the two discounts off the original price.

Use at least one model to solve each problem (e.g., tape diagram, table, double number line diagram, 10×10 grid).

a.   Make a prediction. Do you think the claim is true or false? _____ Explain.

b.   Sam purchased 3 games for $140 after a discount of 30%. What was the original price?

EUREKA
MATH

c. If Sam had used a 20% off coupon and opened a frequent shopper discount membership to save 10%, would the games still have a total of $140?

d. Do you agree with the claim? _____ Explain why or why not. Create a new example to help support your claim.

> **Lesson Summary**
>
> Percent problems have three parts: whole, part, percent.
>
> Percent problems can be solved using models such as ratio tables, tape diagrams, double number line diagrams, and 10×10 grids.

Name _____   Date _____

Angelina received two discounts on a $50 pair of shoes. The discounts were taken off one after the other. If she paid $30 for the shoes, what was the percent discount for each coupon? Is there only one answer to this question?

Angelina received two discounts on a $50 pair of shoes. The discounts were taken off one after the other. If she paid $30 for the shoes, what was the percent discount for each coupon? Is there only one answer to this question?

1. Tony completed filling 12 out of a total of 15 party bags for his little sister's party. What percent of the bags does Tony still have to fill?

   20% *of the bags still need to be filled*.

   > Since there are a total of 15 bags and Tony already filled 12, he has 3 bags left to fill. Now I have to find out what percent 3 out of a total of 15 is. I will write a fraction $\frac{3}{15}$ and rename the fraction as $\frac{1}{5}$. I know $\frac{1}{5}$ is equal to $\frac{20}{100}$. So, Tony has to fill 20% of the bags.

2. Amanda got a 95% on her math test. She answered 19 questions correctly. How many questions were on the test?

   There were **20 *questions on the test*.**

   > I know 19 is a part, and we have to find the total. I will write 95% as a fraction and find an equivalent fraction where 19 is the part. $95\% = \frac{95}{100} = \frac{19}{20}$. $95 \div 5 = 19$, and $100 \div 5 = 20$.

3. Nate read 40% of his book containing 220 pages. What page did he just finish?

   ***Nate just finished page* 88.**

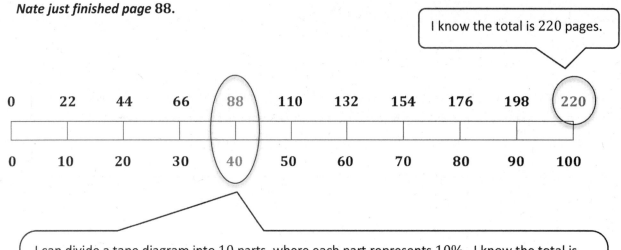

I know the total is 220 pages.

I can divide a tape diagram into 10 parts, where each part represents 10%. I know the total is 220, which represents 100%, so I will divide 220 by 10 to determine the value of 10%. 10% is equal to 22. I will use this information to determine equivalent fractions and label the rest of my tape diagram. I know 40% is 88, so Nate just finished page 88.

Lesson 29:   Solving Percent Problems

EUREKA
MATH

1. Henry has 15 lawns mowed out of a total of 60 lawns. What percent of the lawns does Henry still have to mow?

2. Marissa got an 85% on her math quiz. She had 34 questions correct. How many questions were on the quiz?

3. Lucas read 30% of his book containing 480 pages. What page is he going to read next?

Credits

Great Minds® has made every effort to obtain permission for the reprinting of all copyrighted material. If any owner of copyrighted material is not acknowledged herein, please contact Great Minds for proper acknowledgment in all future editions and reprints of this module.